f-Block Chemistry

f-Block Chemistry

Prof. Helen C. Aspinall

Department of Chemistry, University of Liverpool

OXFORD

UNIVERSITY PRESS

UNIVERSITY PRESS

Great Clarendon Street, Oxford, OX2 6DP,
United Kingdom

Oxford University Press is a department of the University of Oxford.
It furthers the University's objective of excellence in research, scholarship,
and education by publishing worldwide. Oxford is a registered trade mark of
Oxford University Press in the UK and in certain other countries

Published in the United States of America by Oxford University Press
198 Madison Avenue, New York, NY 10016, United States of America

British Library Cataloguing in Publication Data
Data available

Library of Congress Control Number: 2019956869

ISBN 978-0-19-882518-0

Printed in Great Britain by
Bell & Bain Ltd., Glasgow

Contents

Introducing the f-elements

1.1 Introduction

lanthanoids	La	Ce	Pr	Nd	Pm	Sm	Eu	Gd	Tb	Dy	Ho	Er	Tm	Yb	Lu
actinoids	Ac	Th	Pa	U	Np	Pu	Am	Cm	Bk	Cf	Es	Fm	Md	No	Lr

This book is about the elements La to Lu and Ac to Lr, which are referred to using various terms that are almost, but not quite, equivalent. IUPAC has defined the lanthanoids (represented in this book as Ln) as the 15 elements from La to Lu; this definition recognizes that La (electron configuration $[Xe]5d^1 6s^2$) is in many respects a prototype for the elements Ce to Lu in which the 4f orbitals are progressively filled. Similarly, the actinoids (represented in this book as An) are defined by IUPAC as the elements from Ac to Lr. There is still (2018) debate around which elements should appear in Group 3 of the Periodic Table (Sc, Y, La and Ac or Sc, Y, Lu, Lr), with the remaining lanthanoids and actinoids being placed at the bottom of the Table. The commonly used term 'rare earth elements' has historical origins and includes all of the lanthanoids along with Y and Sc. The term 'lanthanides' is still in use, and refers to the elements Ce to Lu; 'actinides' refers to Th to Lr. The lanthanoids and actinoids together are frequently referred to as the 'f-elements' in recognition of the filling of the 4f and 5f orbitals. The chemistry of Y has many similarities with that of the lanthanoids (and Y occurs in nature alongside the lanthanoids) and so it will make occasional appearances in this book.

Y, Tb, Er, and Yb are all named after the small Swedish village of Ytterby near Stockholm.

U and Th, the only two actinoids that occur naturally in significant quantitites, were the first f-elements to be isolated (U in 1789 and Th in 1829), pre-dating all of the lanthanoids, which were discovered between 1839 (Ce) and 1945 (Pm). Detailed investigations of f-element chemistry did not begin until the mid-twentieth century: this was because of the technical difficulties of separation and isolation of the elements in sufficient quantities to study. Spectroscopy was key to the development of rare earth chemistry: it was the most reliable method of determining purity. For example, for many years, 'didymium' (a mixture of mainly Pr and Nd) was considered to be a single element, but careful examination of absorption spectra of didymium isolated from different ores showed small differences in the relative intensities of the electronic absorption bands, suggesting that didymium was in fact a mixture of at least two elements. Carl Auer von Welsbach isolated neodymium ('new twin') and praseodymium ('green twin') from didymium in 1885. Separation and isolation of the elements will be covered in Chapter 7.

The first commercial application of the f-elements was the use of Ce-doped ThO_2 in the gas mantle by Carl Auer von Welsbach in 1891, an invention that made gas lighting practicable. Von Welsbach also invented a useful means of lighting a gas flame when he discovered mischmetal (or pyrophoric alloy), a mixture of mainly La, Ce, and other lanthanoids blended with iron oxide and magnesium oxide that is used in lighter flints. Further applications of the f-elements (e.g. lanthanoids in permanent magnets, phosphors, battery alloys, and catalysts, and uranium in nuclear reactors) were made possible by the development of practicable isolation and separation techniques for the elements during the twentieth century. Several of the lanthanoids (Nd, Eu, Tb, and Dy) are now identified as strategically important metals, particularly in energy applications, and there is concern about possible risks to supply. Applications of f-elements will be covered in Chapter 6.

Investigations of actinoid chemistry are fraught with practical difficulties, and as a consequence actinoid chemistry is nowhere near as well-studied as lanthanoid chemistry. Only two actinoid elements—Th and U—occur naturally to any extent, so most actinoids are not available in the quantities necessary for chemical studies using standard techniques. All isotopes of the actinoids are radioactive, posing safety risks in their handling, and for isotopes with short half-lives, radioactive decay to other elements is likely to occur during the timescale of a typical experiment. Radioactivity may also cause unwanted side-reactions, e.g. radiolytic decomposition of water to H_2O_2 and H_2. To overcome some of these difficulties, special techniques have been developed for studying actinoid chemistry. Tracer techniques use a very low concentration (approx. 10^{-12} M) of An doped into a non-radioactive analogue. Radioactivity of the products is monitored as the reaction proceeds and the fate of the An element can be traced. Ultramicrochemical techniques use very small volumes of solutions at 'normal' concentrations (10^{-1} to 10^{-3} M of An) and manipulations are performed under a microscope.

1.2 Abundance and distribution of the elements

1.2.1 Lanthanoids and Y

All of the lanthanoids (apart from Pm) and Y have crustal abundances that exceed those of Hg and I (Figure 1.1). Pm ($Z = 61$) has no stable isotopes: all of its isotopes decay to the neighbouring elements Nd ($Z = 60$) or Sm ($Z = 62$), and so Pm does not occur naturally to any significant extent.

Because the lanthanoids and Y have very similar chemistry, they occur together in nature. They are all lithophiles and the main ores are monazite ($LnPO_4$), bastnaesite ($LnCO_3F$), and xenotime (YPO_4). Although there is some fractionation of early Ln (La to Eu) and late Ln (Gd to Lu, and Y) between ores, all Ln ores contain mixtures of the elements; the technological challenge of separating the elements will be covered in Chapter 7.

Rare earth ores are widely distributed across the globe, and China is estimated to have the largest accessible deposits of the elements as shown in Figure 1.2.

Pm was first isolated at Oak Ridge National Laboratory in 1945 as a fission product from U.

1.2.2 **Actinoids**

The only two actinoids that occur naturally to any significant extent are U and Th; both these elements are more abundant than As, and Th is more abundant than Sn. U and Th have distinctly different chemistries and so these two elements occur separately in nature: Th mainly occurs alongside early Ln in monazite, and

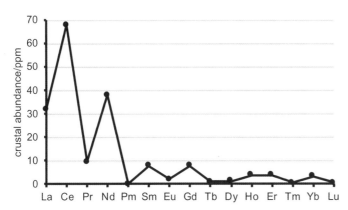

Figure 1.1 Crustal abundance of Ln elements

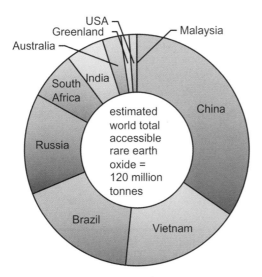

Figure 1.2 Estimated accessible rare earth deposits (data from U.S. Geological Survey 2017)

the main ore for U is uraninite (also known as pitchblende) which consists mainly of UO_2 with variable amounts of U_3O_8 due to oxidation. U is widely distributed, and the largest reserves are in Australia as shown in Figure 1.3.

Uranium occurs naturally as two isotopes: ^{238}U (99.3%) and ^{235}U (0.7%). Both of these are radioactive and they decay to form ultimately ^{206}Pb and ^{207}Pb as shown in Figures 1.4 and 1.5. The ^{238}U series is referred to as the '4n+2' series because the mass numbers of all species have values of 4n+2. In the same way, the ^{235}U decay series is referred to as the '4n+3' series.

The relative abundance of ^{235}U ($t_{1/2} = 7.04 \times 10^8$ y) compared with ^{238}U ($t_{1/2} = 4.46 \times 10^9$ y) is diminishing with time.

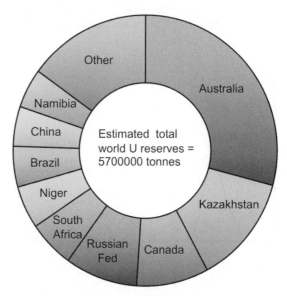

Figure 1.3 Estimated world reserves of U (data from *Uranium 2016: Resources, Production and Demand 'Red Book'*)

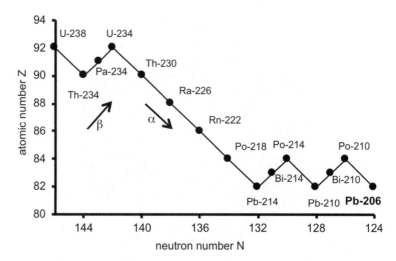

Figure 1.4 The 4n+2 decay series

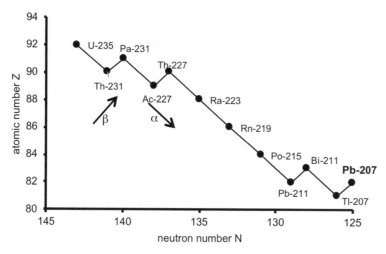

Figure 1.5 The 4n+3 decay series

Pa and Ac occur in very small quantities from radioactive decay of ^{235}U, and all trans-uranium elements are synthetic. The largest single quantity of Pa ever to be collected together was 130 g that was isolated from repeated extractions of 60 tonnes of pitchblende in the 1960s. Approximately 70 tonnes of reactor grade Pu is produced annually in spent nuclear fuel.

The main source of actinoids other than Th and U is neutron irradiation of lighter elements in nuclear reactors as shown in the schemes below:

$$^{226}_{88}Ra + {}^{1}_{0}n \rightarrow {}^{227}_{88}Ra + \gamma \xrightarrow{-\beta^-} {}^{227}_{89}Ac$$

$$^{230}_{90}Th + {}^{1}_{0}n \rightarrow {}^{231}_{90}Th + \gamma \xrightarrow{-\beta^-} {}^{231}_{91}Pa$$

$$^{235}_{92}U + {}^{1}_{0}n \rightarrow {}^{236}_{92}U + \gamma \xrightarrow{{}^{1}_{0}n} {}^{237}_{92}U + \gamma \xrightarrow{-\beta^-} {}^{237}_{93}Np$$

$$^{238}_{92}U + {}^{2}_{1}D \rightarrow {}^{238}_{93}Np + 2{}^{1}_{0}n \xrightarrow{-\beta^-} {}^{238}_{94}Pu$$

$$^{238}_{92}U + {}^{1}_{0}n \rightarrow {}^{239}_{92}U \xrightarrow{-\beta^-} {}^{239}_{93}Np \xrightarrow{-\beta^-} {}^{239}_{94}Pu$$

$$^{238}_{92}U + 17{}^{1}_{0}n \rightarrow {}^{255}_{92}U \xrightarrow{-\beta} {}^{255}_{93}Np \xrightarrow{-\beta} {}^{255}_{94}Pu \xrightarrow{-\beta} \cdots\cdots \xrightarrow{-\beta} {}^{255}_{100}Fm$$

The elements Am to Fm are usually obtained as by-products of the large-scale production of ^{239}Pu.

$$^{239}_{94}Pu + 2{}^{1}_{0}n \rightarrow {}^{241}_{94}Pu \xrightarrow{-\beta^-} {}^{241}_{95}Am$$

Formation of trans-plutonium elements by slow neutron capture requires very high neutron fluxes and often prolonged times: for example irradiation of 1 kg

Pu for 5–10 years at a neutron flux of 3×10^{14} cm^{-2} s^{-1} would produce approximately 1 mg of Cf. Several of these elements are only available in mg quantities or less per year, and investigation of their chemistry is only made possible by the use of tracer techniques.

1.3 The elements

All of the f-elements are very electropositive and tarnish rapidly in air. The lanthanoid metals generally show hexagonal close packed or cubic close packed structures at room temperature, and body centred cubic structures at high temperatures (Eu is anomalous and displays body centred cubic structure under ambient conditions). The structures of the actinoid metals are much more complex, especially for the early part of the series, where Pu has six allotropic modifications, and U and Np have three. The elements Th to Pu all adopt a body centred cubic (bcc) structure at the melting point; Am to Es adopt a face centred cubic (fcc) structure at the melting point and double hexagonal close packed at lower temperatures. The complexity of the structures of the early An is due to the increased radial extent of the 5f orbitals in these elements, which allows them to affect interatomic interactions in the solid state.

1.4 Relativity and the f-elements

We normally associate relativistic effects with systems on an astronomical scale, but they also apply to the atomic scale, and particularly to heavy atoms such as the actinoids.

As the nuclear charge increases, so electrons (particularly s electrons, which have no node at the nucleus) must travel at increased speeds to avoid capture by the nucleus. For heavier atoms such as the actinoids, this speed can approach (but not exceed) the speed of light. According to Einstein's theory of special relativity the electron mass (m) increases according to:

$$m = \frac{m_0}{\sqrt{(1-(v/c))}}$$

where m_0 = resting mass, v = velocity and c = speed of light.

As the mass of the electron increases, the Bohr radius a_0 decreases as:

$$a_0 = (4\pi\varepsilon_0)\left(\hbar^2/me^2\right)$$

where ε_0 = permittivity of a vacuum, $\hbar$ = reduced Planck's constant ($h/2\pi$) and e = charge on an electron.

Relativistic effects primarily apply to 1s electrons, which are very close to the nucleus, but other s orbitals are also affected as they must remain orthogonal to each other, so contraction of the 1s orbital results in contraction and stabilization of all s orbitals in order to maintain orthogonality. Most p orbitals also contract and stabilize. As the s- and p-orbitals contract, they become more effective at screening the d- and f-orbitals from the nuclear charge and so the d- and f-orbitals expand and are destabilized. The expansion of d- and f-orbitals due to relativity is particularly important for the actinoids and has a real impact on the chemistry of the elements: e.g. the 5f electrons of the actinoids are more weakly bound (and hence more chemically active) than the 4f electrons of the lanthanoids.

Spin-orbit coupling—the interaction of an electron's spin angular momentum with its orbital angular momentum—is also a relativistic effect. Its magnitude increases with increasing atomic number and so is much larger for the actinoids than for the lanthanoids (see Chapter 2).

1.5 f-orbitals

The progressive filling of the 4f and 5f orbitals is the characteristic feature of the Ln and An series, so an appreciation of the shapes and radial distribution functions of the f orbitals is important for an understanding of f-element chemistry.

1.5.1 Shapes of the f-orbitals

The seven f-orbitals have m_l values of $0, \pm1, \pm2,$ and ±3. There are different ways of choosing a set of f-orbitals, depending on the symmetry of the species of interest. One set (the 'general set') is shown in Figure 1.6. Their labels are: z^3 ($m_l = 0$); xz^2, yz^2 ($m_l = \pm1$); $xyz, z(x^2 - y^2)$ ($m_l = \pm2$); $x(x^2 - 3y^2), y(3x^2 - y^2)$ ($m_l = \pm3$). Figure 1.7 shows views along the z axis illustrating the angular nodes of these orbitals.

For convenience, a different set of f-orbitals is chosen for systems with cubic (O_h or T_d) symmetry. These orbitals are labelled: xyz (A_{2u}); x^3, y^3, z^3 (T_{1u}); $z(x^2 - y^2),$ $x(z^2 - y^2), y(z^2 - x^2)$ (T_{2u}).

In a crystal field of O_h symmetry, the cubic set of f-orbitals split into A_{2u} (xyz); T_{1u} (x^3, y^3, z^3); T_{2u} ($z(x^2 - y^2),$ $x(z^2 - y^2),$ $y(z^2 - x^2)$).

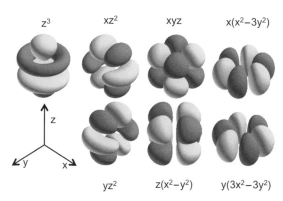

Figure 1.6 The general set of f-orbitals (D.L. Cooper, University of Liverpool)

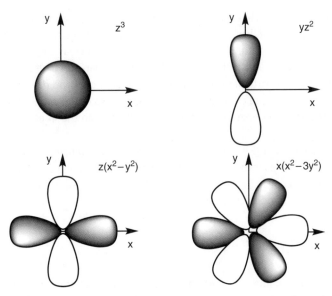

Figure 1.7 f-orbitals viewed along the z axis to illustrate the angular nodes: z^3: 0 nodes; xz^2, yz^2: 1 node; $z(x^2 - y^2)$, xyz: 2 nodes; $x(x^2 - 3y^2)$; $y(3x^2 - y^2)$: 3 nodes

1.5.2 **Radial distribution functions**

Radial distribution functions for the valence orbitals of Sm^{3+} and its actinoid analogue Pu^{3+} are shown in Figure 1.8. The 4f orbitals of Sm^{3+} (and of other Ln^{3+}) are essentially core orbitals and have very little interaction with any surrounding

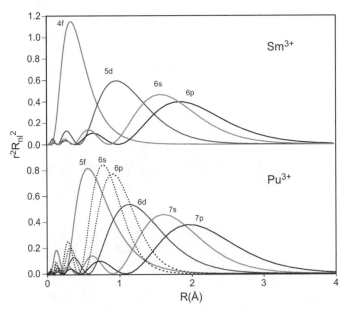

Figure 1.8 Relativistic radial distribution functions for Sm^{3+}(6-coord. radius 1.10 Å) and Pu^{3+}(6-coord. radius 1.14 Å). Reproduced with permission from M. L. Neidig, *Coord. Chem. Rev.*, **257**, 394. Copyright 2013 Elsevier.

crystal field. The impact of this on spectroscopic and magnetic properties will be seen in Chapter 2. Relativistic effects (Section 1.4) result in the 5f orbitals of Pu^{3+} having a much greater radial extent than the 4f orbitals of Sm^{3+}, and there is the potential for the 5f orbitals of the early An to take part in some covalent bonding with ligands. This difference between 4f and 5f orbitals can result in distinct differences in coordination and organometallic chemistry between the early An and their Ln analogues. (The 6d orbitals of early An are also able to take part in covalent bonding.)

1.6 Electron configurations

The lanthanoids and actinoids can be considered as 4f and 5f analogues of the d-transition metals. This view is supported by the electron configurations of the gas phase atoms and ions summarized in Table 1.1.

1.6.1 Electron configurations for Ln

Electron configurations of the lanthanoid gas phase atoms and their 3+ ions are given in Table 1.1. The filling of the 4f orbitals follows a fairly regular pattern along the series; the irregularity at Gd ($[Xe]4f^75d^16s^2$) reflects the stability of the half-filled 4f shell. In the Ln^{3+} ions the 4f orbitals are progressively filled from La^{3+} ($4f^0$) to Lu^{3+} ($4f^{14}$), and the stability of the $4f^7$ and $4f^{14}$ configurations is reflected

Table 1.1 Electron configurations of lanthanoid (and Y) and actinoid atoms and ions

Element	Atom	Ln^{3+}	Element	Atom	$An^{2+}(g)$	$An^{3+}(g)$
La	$[Xe]5d^16s^2$	$[Xe]$	Ac	$[Rn]6d^17s^2$	$[Rn]7s^1$	
Ce	$[Xe]4f^15d^16s^2$	$[Xe]4f^1$	Th	$[Rn]6d^27s^2$	$[Rn]5f^16d^1$	$[Rn]5f^1$
Pr	$[Xe]4f^36s^2$	$[Xe]4f^2$	Pa	$[Rn]5f^26d^17s^2$	$[Rn]5f^26d^1$	$[Rn]5f^2$
Nd	$[Xe]4f^46s^2$	$[Xe]4f^3$	U	$[Rn]5f^36d^17s^2$		$[Rn]5f^3$
Pm	$[Xe]4f^56s^2$	$[Xe]4f^4$	Np	$[Rn]5f^46d^17s^2$		$[Rn]5f^4$
Sm	$[Xe]4f^66s^2$	$[Xe]4f^5$	Pu	$[Rn]5f^67s^2$	$[Rn]5f^6$	$[Rn]5f^5$
Eu	$[Xe]4f^76s^2$	$[Xe]4f^6$	Am	$[Rn]5f^77s^2$	$[Rn]5f^7$	$[Rn]5f^6$
Gd	$[Xe]4f^75d6s^2$	$[Xe]4f^7$	Cm	$[Rn]5f^76d7s^2$	$[Rn]5f^8$	$[Rn]5f^7$
Tb	$[Xe]4f^96s^2$	$[Xe]4f^8$	Bk	$[Rn]5f^97s^2$	$[Rn]5f^9$	$[Rn]5f^8$
Dy	$[Xe]4f^{10}6s^2$	$[Xe]4f^9$	Cf	$[Rn]5f^{10}7s^2$	$[Rn]5f^{10}$	$[Rn]5f^9$
Ho	$[Xe]4f^{11}6s^2$	$[Xe]4f^{10}$	Es	$[Rn]5f^{11}7s^2$	$[Rn]5f^{11}$	$[Rn]5f^{10}$
Er	$[Xe]4f^{12}6s^2$	$[Xe]4f^{11}$	Fm	$[Rn]5f^{12}7s^2$		$[Rn]5f^{11}$
Tm	$[Xe]4f^{13}6s^2$	$[Xe]4f^{12}$	Md	$[Rn]5f^{13}7s^2$		$[Rn]5f^{12}$
Yb	$[Xe]4f^{14}6s^2$	$[Xe]4f^{13}$	No	$[Rn]5f^{14}7s^2$		$[Rn]5f^{13}$
Lu	$[Xe]4f^{14}5d^16s^2$	$[Xe]4f^{14}$	Lr	$[Rn]5f^{14}7s^27p^1$		$[Rn]5f^{14}$

in the accessibility of the Eu^{2+} ($4f^7$) and Yb^{2+} ($4f^{14}$) oxidation states, and in the anomalously large metallic radii for Eu and Yb (see Section 1.7).

1.6.2 Electron configurations for An

Actinoid electron configurations are shown in Table 1.1, alongside those of their lanthanoid analogues. There is a striking contrast between the configurations of early An (Th to Np) and their 4f congeners: the 6d orbitals of Th to Np are very close in energy to the 5f orbitals (lower in Th) resulting in 6d orbital occupation in the gas phase atoms. From Pu to No, as the 5f orbitals become lower in energy compared with 6d, the gas phase An and An^{3+} follow the same pattern as the lanthanoids (see Figure 1.9). However, Lr is anomalous ($5f^{14}7s^27p^1$): this is due to the stabilization of 7p cf 6d by relativistic effects. The contrast in electron configurations between early An and early Ln is mirrored in contrasting chemistry of these elements, which will be a recurring theme throughout this book.

1.7 Metallic and ionic radii

Metallic (12-coordinate) and selected ionic (6-coordinate) radii for Ln and An are shown in Figures 1.10 and 1.11. Data for the actinoids are much less complete than for the lanthanoids, due to practical difficulties associated with obtaining sufficient quantities of materials to characterize by x-ray crystallography.

1.7.1 Metallic radii

Metallic radii are obtained from x-ray crystal structure determination of the metals, and to take account of different crystal structures and coordination numbers (particularly for An), the radii plotted in Figures 1.10 and 1.11 have all been corrected to 12-coordinate radii using atomic volume data.

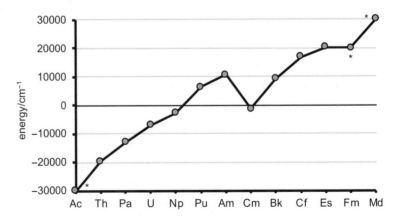

Figure 1.9 Relative energies of $5f^{n-1}6d7s^2$ configuration compared with $5f^n7s^2$ for gas phase An atoms. * = estimated.

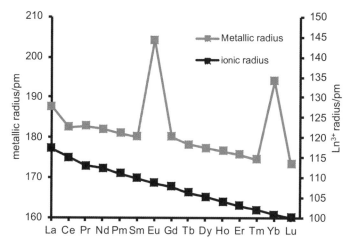

Figure 1.10 Metallic and Ln^{3+} radii for Ln. Ln^{3+} values are 6-coordinate (Shannon)

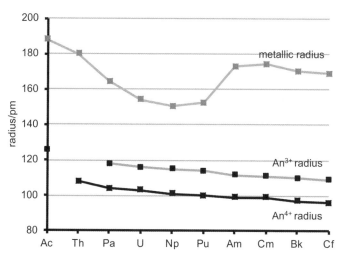

Figure 1.11 Metallic and ionic radii for An. Ionic radii are 6-coordinate (Shannon)

Two features of Ln metallic radii are immediately apparent: firstly, there is a general decrease from La to Lu, and secondly, the radii of Eu and Yb are anomalously large.

The anomalous radii of metallic Eu and Yb are a consequence of the stability of half-filled and filled 4f sub-shells (this will be a recurring theme throughout the chemistry of Eu and Yb). The electron configurations of metallic Eu and Yb are [Xe]4f^{7}6s^2 and [Xe]4f^{14}6s^2 respectively, leaving just the 6s^2 electrons available for delocalized metallic bonding (the 4f orbitals of Ln are essentially core orbitals and do not take part in bonding). Metallic Eu and Yb can be thought of as Ln^{2+} ions in a delocalized sea of bonding electrons. By contrast, other metallic Ln have a [Xe]4f^{n}5d^{1}6s^2 configuration in which three electrons per atom contribute

Anomalously large metallic radii for Eu and Yb are mirrored by anomalously low boiling points (1870 K and 1466 K respectively) compared with other Ln (e.g. La 3730 K and Lu 3668 K).

to bonding so that these metals consist of Ln^{3+} ions in a sea of delocalized bonding electrons. The radii of Ln^{2+} are larger than those of Ln^{3+} resulting in a larger metallic radius for Eu and Yb than for other Ln. The metallic radius of Ce is very slightly irregular and has been the subject of numerous studies. The 4f orbitals are not fully stabilized in Ce and there is a possibility of 4f orbital contribution to bonding.

The metallic radii of the actinoids (Figure 1.11) are much less regular than those of the lanthanoids (Figure 1.10). The early actinoids Ac to Np show a parabolic decrease in metallic radius that resembles closely the decrease in radius for 5d transition metals. This observation is consistent with 5f orbital contributions to metallic bonding for the early actinoids. For the later actinoids from Am, the metallic radii resemble those of the lanthanoids, consistent with the 5f orbitals becoming core-like and making no contribution to bonding.

1.7.2 Ionic radii

The 6-coordinate Ln^{3+} radii (Figure 1.10; from Shannon 1976) show a very clear and steady decrease of approximately 14% from La^{3+} to Lu^{3+} as the 4f orbitals are progressively filled. Similarly, $An^{3+/4+}$ radii (Figure 1.11) show a steady decrease from Ac to Cf with the progressive filling of the 5f orbitals. An^{3+} radii show a greater decrease than those of corresponding Ln^{3+}: the decrease from Ac^{3+} to Cf^{3+} is 17 pm (13.4% of the Ac^{3+} radius) whereas the decrease from La^{3+} to Dy^{3+} is 12.2 pm (10.4% of the La^{3+} radius).

These trends in ionic radii have a profound influence on the coordination chemistry of the lanthanoids and actinoids (e.g. coordination numbers, Brønsted acidity of aqua ions), which will be encountered in numerous examples throughout this book.

1.7.3 Lanthanoid and actinoid contractions

The decreases in ionic and metallic radii for Ln and An with increasing atomic number are referred to as the lanthanoid and actinoid contractions. The lanthanoid contraction not only influences Ln coordination chemistry: it is also responsible for the very similar radii of 4d and 5d transition elements within the same group (e.g. 6-coordinate radii for Zr^{4+} and Hf^{4+} are 86 pm and 85 pm respectively).

The lanthanoid and actinoid contractions arise in part due to the poor screening ability of the 4f and 5f orbitals. As atomic number increases and electrons are added to the f sub-shells, the valence electrons are not effectively screened from the increased nuclear charge, resulting in a decreased radius. However, this electrostatic effect cannot explain the full magnitude of the lanthanoid and actinoid contractions, and relativistic effects must be considered in a more complete explanation. (See Section 1.4.)

Relativistic contraction of inner s orbitals results in *relativistic expansion* of f orbitals.

It is estimated that 10–30% of the lanthanoid contraction and 40–50% of the actinoid contraction are due to relativistic effects.

1.8 Ionization potentials and oxidation states for Ln

1.8.1 Ionization potentials

Figure 1.12 shows the first three ionization potentials (IPs) for Ln, and Figure 1.13 shows the sum of the first four IPs. The first and second IPs vary very little along the series; however, the third IPs show distinct maxima at Eu and Yb. This is due to the extra stability associated with half-filled and filled 4f sub-shells for Eu^{2+} ($4f^7$) and Yb^{2+} ($4f^{14}$).

> The extra stability of a half-filled sub-shell with parallel spins is due to a quantum mechanical phenomenon known as 'exchange stabilization'.

In all cases the fourth IP, which for Ce to Lu requires removal of an electron from the 4f sub-shell, is larger than the sum of the first three IPs, explaining the predominance of the +3 oxidation state throughout Ln chemistry.

The fourth IP (and the sum of the first four IPs) is at a minimum for Ce. Formation of Ce^{4+} requires removal of the single 4f electron from Ce^{3+} ([Xe]$4f^1$) to leave a stable closed shell configuration. As a consequence Ce is the only lanthanoid that has an accessible +4 oxidation state in solution. Pr^{4+} ($4f^1$) and Tb^{4+} ($4f^7$) are both accessible in the solid state (see Chapter 3 for examples of the +4 oxidation state in oxides), but are too powerfully oxidizing to exist in solution.

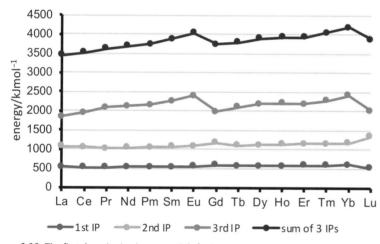

Figure 1.12 The first three ionization potentials for Ln

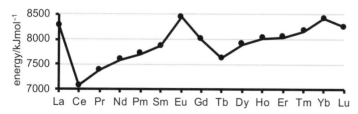

Figure 1.13 Sum of first four ionization potentials for Ln

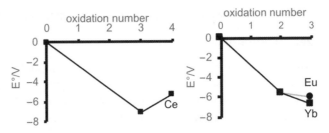

Figure 1.14 Frost diagrams for Ce, Eu, and Yb

1.8.2 **Oxidation states**

The Frost diagrams in Figure 1.14 give a graphical representation of experimentally determined redox potentials that are relevant to chemistry in aqueous solution. For all Ln in aqueous solution the +3 oxidation state is the most stable, but Ce^{4+} and Eu^{2+} can exist.

Other oxidation states are accessible in the solid state or, with the correct choice of ligand, in non-aqueous solution. Examples are given in Chapters 4, 5, and 6.

Lanthanoid oxidation states: summary of main points

- The +3 oxidation state is in general the most stable for all lanthanoids.
- The +2 oxidation state is most accessible for Eu^{2+} $(4f^7)$ and Yb^{2+} $(4f^{14})$ both of which can exist in solution.
- The +4 oxidation state is most accessible for Ce^{4+} $(4f^0)$, which can exist in solution.
- The +4 oxidation state is accessible in the solid state for Pr^{4+} $(4f^1)$ and Tb^{4+} $(4f^7)$.

1.9 **Ionization potentials and oxidation states for An**

1.9.1 **Ionization potentials**

Experimental data for An IPs are far less complete than those for Ln (only Th and U have experimental values for more than one IP). However, first IPs have been measured experimentally for all An except Pa, Fm, Md, and No, and are plotted in Figure 1.15. The very similar energies of 5f and 6d orbitals for the early An result in irregular ground state electron configurations for these elements (Table 1.1), and corresponding irregularities in values for first IPs. For the later An from Pu to No, electron configurations correspond with those of the Ln analogues, and are mirrored by trends in 1st IPs. In Lr, the heaviest actinoid, relativistic effects lead to stabilization of the $7p_{1/2}$ orbital with respect to 6d, resulting in a ground state $[Rn]5f^{14}7s^27p^1$ configuration. The 7p electron is very weakly bound, and consequently Lr has the lowest 1st IP of all the f-elements.

The 1st IP for ^{256}Lr $(t_{1/2} = 27$ s$)$ was reported in 2015

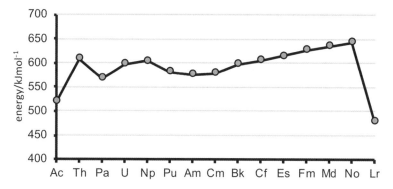

Figure 1.15 1st IPs for An. Values for Pa, Fm, Md, and No are estimated

1.9.2 **Oxidation states**

Figure 1.16 summarizes accessible oxidation states and 5f configurations for Ac to Cm, and the relative stabilities of oxidation states are summarized in the Frost diagrams in Figure 1.17. From Cm to Lr, the oxidation states follow a similar pattern to those of the Ln series, and (apart from No) the +3 state is the most stable. For No, the +2 oxidation state (No^{2+} $5f^{14}$) is the most stable due to the full 5f sub-shell.

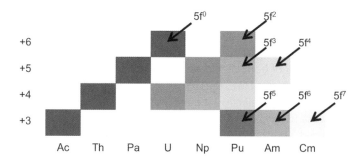

Figure 1.16 Summary of important oxidation states and 5f configurations for selected An

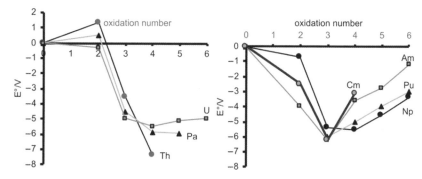

Figure 1.17 Frost diagrams for selected An

Actinoid oxidation states: summary of main points

- Ac to U all have a stable oxidation state with $5f^0$ configuration.
- Pa to Am all have several accessible oxidation states.
- The +3 oxidation state is the most stable for Ac, and Am to Lr (except No).
- The +4 oxidation state is most stable for Th.
- The +2 oxidation state is the most stable for No (No^{2+} $5f^{14}$).

1.10 Summary

All of the lanthanoid elements (apart from Pm) occur naturally in greater abundance than either iodine or mercury, but they always occur as mixtures in their ores due to the great similarity in their properties. The similarity in properties means that they cannot be separated by purely chemical means and this proved a major challenge for the isolation of the individual lanthanoids. All of the actinoids are radioactive, and only uranium and thorium occur naturally to any significant extent. The trans-uranium elements are all synthetic and are formed by neutron capture.

The characteristic feature of lanthanoids and actinoids is the progressive filling of the 4f and 5f orbitals respectively. The f-orbitals of the lanthanoids and the later actinoids behave as core orbitals and do not take part in bonding; however, the greater radial extent of the 5f orbitals of the early actinoids means that they can contribute to bonding. The similarity in energies of 5f and 6d orbitals in the early An results in distinctive chemistry for these elements, and several oxidation states are accessible for Pa to Am. For Ln and the later An, the +3 oxidation state is generally the most stable.

Metallic and ionic radii of both Ln and An decrease with increasing atomic number; these effects are referred to as the lanthanoid and actinoid contractions and are due to a combination of electrostatic and relativistic effects. In f-element chemistry, the decrease in radius has an influence on coordination chemistry as the series are traversed, and the lanthanoid contraction is responsible for the similarity in radii between 2nd and 3rd row d-transition metals and their ions.

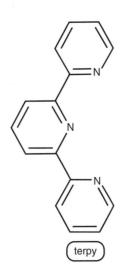

terpy

1.11 Exercises

1. The present-day ratio of ^{238}U: ^{235}U is 99.3 : 0.7. Calculate what this ratio will be in 1 billion years' time. $t_{1/2}(^{238}U) = 4.46 \times 10^9$ y; $t_{1/2}(^{235}U) = 7.04 \times 10^8$ y.
2. Explain the trend in average Ln–N distances for the 10-coordinate complexes [Ln(terpy) $(NO_3)_3(H_2O)$]:

Ln	Ce	Pr	Nd	Sm	Eu	Tb	Dy	Ho
average Ln–N distance/pm	264.1	262.1	260.4	258.1	255.4	254.9	252.8	252.3

Data from *Coordination Chemistry Reviews*, 2017, **340**: 220.

3. Predict what will be formed when each of the following metals is dissolved in dilute hydrochloric acid: Ce, Th, Pa, U, Pu.

2 Electronic structure, magnetism, and spectroscopy

2.1 Introduction

The majority of f-element atoms and ions have unpaired f-electrons, and many of the unique properties of the elements and their compounds arise from the way in which these electrons interact with each other. This chapter will show how Hund's rules can be used to work out the most stable arrangement of electrons in the atoms/ions, and how this information can be summarized in a term symbol. We will then look at magnetic properties of the ions and at the effect of paramagnetism on NMR spectra of f-element complexes. The final sections of the chapter will look at electronic absorption spectroscopy and at lanthanoid luminescence. This chapter will focus almost entirely on the lanthanoids as actinoid systems are much more complicated. Magnetism and luminescence are fundamental to some of the most important applications of the lanthanoids (see Chapter 6).

2.2 Deriving term symbols for f-element ions

An electron configuration cannot give all the information about how the electrons are arranged in an atom or ion: for example '$4f^2$' tells us that the 4f orbitals contain a total of two electrons, but it doesn't tell us which of the seven 4f-orbitals are occupied ($m_l = \pm 3, \pm 2, \pm 1, 0$), nor does it give any information about the orientations of the electron spins ($m_s = \pm \frac{1}{2}$). The different ways in which electrons can be arranged are called 'microstates'.

Electrons in a multi-electron atom or ion are not independent of each other: when electrons have orbital angular momentum there are electrostatic repulsions between them, giving rise to quantized microstates that can have different energies. Some of the microstates that arise from electrostatic repulsions can have the same energy and these can be grouped together as *terms*. It is the difference in energy between terms that is measured by spectroscopy.

For lighter atoms and ions (including the lanthanoids) the factors that determine the energy of a microstate (in order of decreasing importance) are:

(i)   orientation of spin angular momentum: the strongest interaction is between electron spins and these couple to give a resultant S that can take the following values:

$$S = s_1 + s_2, \ s_1 + s_2 - 1, \ ..., \ |s_1 - s_2|$$

(ii) orientation of orbital angular momentum: the next strongest interaction is between orbital angular momenta and these couple to give a resultant L that can take the following values:

$$L = l_1 + l_2, \ l_1 + l_2 - 1, \ ..., \ |l_1 - l_2|$$

(iii) spin-orbit coupling: total spin angular momentum (S) couples with total orbital angular momentum (L) to give total angular momentum J that can take the following values:

$$J = |L + S|, \ |L + S - 1|, \ ..., \ |L - S|$$

Spin-orbit coupling is large for the actinoids and so s and l of an individual electron couple to give j and then the j values for individual electrons couple to give a resultant J. This is known as jj coupling.

Spin-orbit coupling is a magnetic-magnetic coupling and therefore much weaker than electrostatic interactions. The magnitude of spin-orbit coupling is determined by the spin-orbit coupling constant, λ, which increases with increasing atomic number Z: $\lambda \approx Z^4$.

This scheme for coupling electron angular momenta is known as the Russell–Saunders coupling scheme and it works reasonably well for lanthanoids and lighter atoms/ions, but not for actinoids. The combination of S and L is known as a *term*, and the combination of S, L, and J is known as a *level*. For the lanthanoids, typical splitting due to electrostatic repulsions is $\approx 10^4$ cm^{-1} and splitting due to spin-orbit coupling is $\approx 10^3$ cm^{-1} (see Figure 2.6 for Pr^{3+} electronic energy levels).

The values of S, L, and J are summarized in a *term symbol* which has the form:

$$^{2S+1}L_J$$

The value of $2S+1$ is referred to as the *multiplicity*; for example $S = 1$ can have $M_s = +1, 0, -1$ and so it is referred to as a 'triplet' state. The value of L is designated by a letter rather than a number as shown below:

L value	0	1	2	3	4	5	6	7
Designation	S	P	D	F	G	H	I	K

S, P, D, and F are derived from spectroscopic lines associated with s, p, d, and f orbitals, and the rest are alphabetical (omitting J).

Once the individual electron spins have been coupled and the individual orbital angular momenta have been coupled, the resultant total spin angular momentum can be coupled (by a process known as spin-orbit coupling) with the total orbital angular momentum to give a resultant angular momentum J, which can take values:

$$J = |L+S|, |L+S|-1, \dots, |L-S|$$

The different J values of a Russell–Saunders term are known as levels. The energy separation Δ_J between adjacent levels is proportional to the larger of the two J values:

$$\Delta_J \approx J(J+1) - (J-1)J = 2J$$

This is known as the 'Landé interval rule'.

2.2.1 Using Hund's rules to determine the ground state term symbol

We are often most interested in the arrangement of electrons in the ground state, and Hund's rules can be used to determine this. Hund's rules state that:

1. The arrangement of electrons with the maximum value of S has the lowest energy. This means that each orbital within the sub-shell is singly occupied before the electrons start to pair. It minimizes the repulsive Coulomb inter-actions between electrons with the same spin.

$$S = \sum m_s$$

2. If there is more than one arrangement of electrons with the maximum value of S, then the arrangement with the maximum value of L has the lowest energy. The maximum value of L occurs when the electrons are all orbiting in the same direction, and this means that they encounter each other least often, thus minimizing Coulomb repulsions.

$$L = \sum m_l$$

3. If the sub-shell is less than half-full then the level with the lowest J value has the lowest energy; if the sub-shell is more than half-full, then the level with the highest J value has the lowest energy. This minimizes the spin-orbit coupling.

$$< \text{half} - \text{filled shell}: J = |L-S|$$

$$> \text{half} - \text{filled shell}: J = L+S$$

Determining the ground state term symbol: examples

1. **Pr^{3+} $4f^2$**

m_l	3	2	1	0	−1	−2	−3
	↑	↑					

The two electrons, both with $m_s = +1/2$, are placed in the two orbitals with the largest m_l values in order to maximize S and L:

Rule 1: maximum $S = \dfrac{1}{2} + \dfrac{1}{2} = 1$

Multiplicity $= 2S + 1 = 3$

Rule 2: maximum $L = 3 + 2 = 5$

$L = 5$ ∴ ground state term is ^3H

Now we need to determine the J value of the lowest level.

Rule 3: the 4f shell is less than half-filled ∴ lowest level has the minimum value of J

Minimum $J = |L - S| = 4$

∴ complete ground state term symbol for Pr^{3+} is $\mathbf{^3H_4}$

2. **Er^{3+} $4f^{11}$**

m_l	3	2	1	0	−1	−2	−3
	↑↓	↑↓	↑↓	↑↓	↑	↑	↑

Because there are 11 electrons to accommodate, four of the seven 4f orbitals must be doubly occupied. In order to maximize L, the four doubly occupied orbitals must be $m_l = 3, 2, 1$, and 0.

Rule 1: maximum $S = \dfrac{1}{2} + \dfrac{1}{2} + \dfrac{1}{2} = \dfrac{3}{2}$

Multiplicity $= 2S + 1 = 4$

Rule 2: maximum $L = 2{\times}3 + 2{\times}2 + 2{\times}1 + 2{\times}0 - 1 - 2 - 3 = 6$

$L = 6$ ∴ ground state term is ^4I

Now we need to determine the J value of the lowest level.

Rule 3: the 4f shell is more than half-filled ∴ lowest level has the maximum value of J

Maximum $J = L + S = \dfrac{15}{2}$

∴ complete ground state term symbol for Er^{3+} is $\mathbf{^4I_{15/2}}$

2.2.2 Effect of crystal field

In most chemically interesting situations, the metal is not a free ion in the gas phase: it is usually surrounded by ligands and so is subject to the influence of a crystal field. Due to the essentially core nature of the 4f orbitals in Ln ions, the interaction between the f-electrons and the crystal field is much smaller than the inter-electronic interactions and the crystal field splitting results in a relatively minor perturbation of the free ion states. The effect of the crystal field is to split each L_J level into $2J+1$ sub-levels with $M_J = J, J-1, \ldots -J$. For systems with an odd number of unpaired electrons (and thus half-integral values of M_J), in the absence of a magnetic field, $\pm M_J$ sub-levels are degenerate. For example, the $^2F_{5/2}$ ground state of Ce^{3+} will be split by a crystal field into three pairs of degenerate sub-levels with $M_J \pm 5/2, \pm 3/2$, and $\pm \frac{1}{2}$, and the $^2F_{7/2}$ state will be split into four degenerate pairs with $M_J \pm 7/2, \pm 5/2, \pm 3/2$, and $\pm \frac{1}{2}$. The relative energies of these sub-levels depend on the exact nature of the crystal field. Systems with an even number of unpaired electrons (even values for M_J) are subject to zero-field splitting of $\pm M_J$ sub-levels, which is often very small. Crystal field splittings for Ln ions are usually at least an order of magnitude smaller than spin-orbit coupling.

In the Ln series, crystal field splitting is *smaller* than spin-orbit splitting; this is a contrast with d-transition metals for which crystal field splitting is *larger* than spin-orbit splitting.

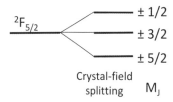

2.3 Magnetic properties

The magnetic properties of an atom, ion, or complex are determined by the total electron angular momentum (i.e. the combination of spin and orbital angular momentum). For a free ion (i.e. a gas phase ion with no ligands) the free ion formula is used to calculate the magnetic moment:

$$\mu_{eff} = g_J \sqrt{J(J+1)}$$

where g_J, the Landé g-factor, is given by:

$$g_J = 1 + \frac{J(J+1) - S(S+1) + L(L+1)}{2J(J+1)}$$

For lanthanoid complexes, the free-ion formula still applies. This is because the 4f orbitals, due to their essentially core nature, interact only very weakly with the ligands and so remain almost degenerate (the crystal field splitting is an order of magnitude smaller than spin-orbit coupling). This means that, as in a free ion, orbital angular momentum is not quenched and so the free ion formula can be applied to complexes.

This model works well for complexes of the lanthanoids, as shown by the data in Table 2.1. The only exceptions are Eu^{3+} and Sm^{3+}, both of which have low-lying excited states that are appreciably populated at room temperature. The magnetic moments of both these ions decrease with decreasing temperature, and Eu^{3+} complexes are diamagnetic as predicted by the free ion formula at low temperatures. When population of excited states is included in the calculation of μ_{eff}, good agreement with experiment is achieved.

2.3.1 Lanthanoid single ion magnets

A 'single molecule magnet' (SMM) is a molecular species that, when placed in an external magnetic field, aligns its magnetic moment along the most energetically

Table 2.1 Ground state term symbols and magnetic properties for Ln^{3+}

Ln	4f configuration	ground state term symbol	g_J	$g_J\sqrt{(J(J+1))}$	experimental μ_{eff} for $[Ln(NO_3)_3(phen)_2]$*
La^{3+}	$4f^0$	$^1S_0$	0	0	0
Ce^{3+}	$4f^1$	$^2F_{5/2}$	6/7	2.54	2.46
Pr^{3+}	$4f^2$	$^3H_4$	4/5	3.58	3.48
Nd^{3+}	$4f^3$	$^4I_{9/2}$	8/11	3.62	3.44
Pm^{3+}	$4f^4$	$^5I_4$	3/5	2.68	
Sm^{3+}	$4f^5$	$^6H_{5/2}$	2/7	0.84	1.64
Eu^{3+}	$4f^6$	$^7F_0$	1	0.0	3.36
Gd^{3+}	$4f^7$	$^8S_{7/2}$	2	7.94	7.97
Tb^{3+}	$4f^8$	$^7F_6$	3/2	9.72	9.81
Dy^{3+}	$4f^9$	$^6H_{15/2}$	4/3	10.63	10.6
Ho^{3+}	$4f^{10}$	$^5I_8$	5/4	10.60	10.7
Er^{3+}	$4f^{11}$	$^4I_{15/2}$	6/5	9.59	9.46
Tm^{3+}	$4f^{12}$	$^3H_6$	7/6	7.57	7.51
Yb^{3+}	$4f^{13}$	$^2F_{7/2}$	8/7	4.54	4.47

*Values from *J. Inorg. Nucl. Chem.* 1965, **27**: pp. 1605–10.

favourable direction (the 'easy axis of magnetization'), and, below a temperature known as the 'blocking temperature', retains its magnetization when the external field is removed. If the SMM contains only one metal ion, then it is known as a 'single ion magnet' (SIM). The first examples of SIMs were lanthanoid bis(phthalocyanine) complexes $[LnPc_2]^-$ (see Chapter 4), reported in 2003, and since then numerous other examples of lanthanoid SIMs have been reported.

The ideal SIM should have a highly anisotropic electron distribution, and the ground state should be split by a crystal field so that the maximum $\pm M_J$ sublevel (corresponding to the largest magnetic moment) has the lowest energy. In order for the SIM to retain its magnetization in the absence of an external magnetic field, there must be a significant energy barrier to relaxation of the magnetic moment. The mechanism for magnetic relaxation is via excited crystal field states, and the stronger the crystal field, the higher in energy these excited states will be.

Ln^{3+} ions that have an oblate electron distribution are stabilized by an axially symmetric crystal field in which the ligand electron density is primarily localized above and below the xy plane, whereas a prolate electron distribution is stabilized by an equatorial ligand field as illustrated in Figure 2.1. In these situations the magnetic moment of the Ln ion is aligned with the z-axis of the crystal field.

The double decker sandwich complexes $[LnPc_2]^-$ have four-fold axial symmetry with the ligand electron density above and below the xy plane as shown

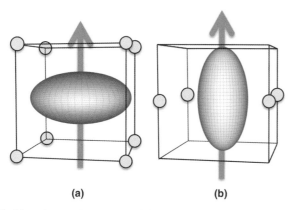

Figure 2.1 (a) Oblate electron distribution stabilized by ligand electron density above and below xy plane. Magnetization (large arrow) aligned with crystal field axis. (b) Prolate electron distribution stabilized by ligand electron density in equatorial (xy) plane. Magnetization (large arrow) aligned with crystal field axis.

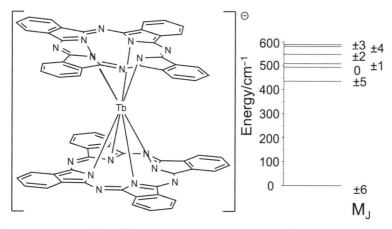

Figure 2.2 Structure of $[TbPc_2]^-$ (left) and crystal field splitting of the $^7F_6$ level in $[TbPc_2]^-$ (right)

in Figure 2.2. Where Ln = Tb, the $^7F_6$ ground state is split by the crystal field so that the $M_J \pm 6$ state, which has oblate electron distribution, is the most stable. The next highest crystal field state ($M_J \pm 5$) is over 400 cm^{-1} higher in energy, and this energy barrier is sufficient to result in slow magnetic relaxation and SIM behaviour at 11.5 K.

[Dy(OBu^t)$_2$(py)$_5$]$^+$ which has approximate D$_{5h}$ symmetry is another example in which a strong axial crystal field results in SIM behaviour with a blocking temperature of 14 K. This complex has short axial Dy–O distances (approx. 2.11 Å) compared with the equatorial Dy–N distances (approx. 2.2 Å). This strongly axial crystal field results in the stabilization of the $M_J \pm 15/2$ state and SIM behaviour at 14 K.

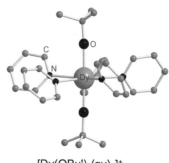

[Dy(OBu^t)$_2$(py)$_5$]$^+$

2.4 NMR spectroscopy of paramagnetic f-element complexes

Almost all complexes of lanthanoids and actinoids are paramagnetic due to the presence of unpaired f-electrons (although Eu^{3+} ($4f^6$) has a diamagnetic $^7F_0$ ground state it has paramagnetic excited states of sufficiently low energy to be significantly occupied at room temperature). When an NMR spectrum is recorded for a paramagnetic complex, the nucleus under investigation experiences the local magnetic field due to the paramagnetic metal ion in addition to the external magnetic field due to the NMR spectrometer. As a result the NMR spectra of paramagnetic complexes are significantly different from those of diamagnetic analogues, and much larger chemical shifts are observed. For most paramagnetic lanthanoid and actinoid complexes, strong spin-orbit coupling results in fast electron spin-lattice relaxation and so their NMR spectra show reasonably sharp lines. Figure 2.3 shows ^1H NMR spectra of the D_3 symmetric binaphtholate complexes $Na_3[Ln(binol)_3]$ for Ln = La ($4f^0$ diamagnetic) and Ln = Yb ($4f^{13}$ paramagnetic). These complexes are described in Chapter 4.

For lanthanoids there is essentially no covalent contribution to metal-to-ligand bonding and thus essentially no delocalization of unpaired f-electrons onto ligands. The mechanism of interaction between ligand nuclei and unpaired f-electrons is therefore almost entirely 'through-space' or dipolar in origin, and requires the metal ion to have an anisotropic distribution of f-electrons. It is often referred to as 'pseudocontact shifting'. If there is a small covalent contribution to bonding, some unpaired f-electron density will be delocalized onto ligand atoms and a 'contact' shift will result.

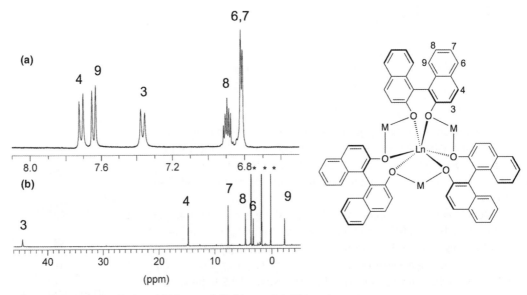

Figure 2.3 ^1H NMR spectra of $Na_3[Ln(binol)_3]$ (a) Ln= La ($4f^0$); (b) Ln = Yb ($4f^{13}$) * = solvent resonances

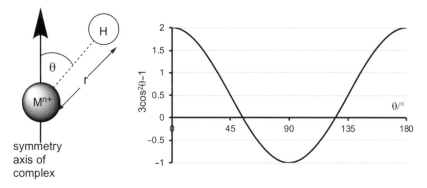

Figure 2.4 Geometrical factors in dipolar shifting for an axially symmetric complex

Pseudocontact shifting cannot occur if the complex has an isotropic magnetic susceptibility. Complexes of Gd^{3+} ($4f^7$), by virtue of their ^8S ground state, have isotropic magnetic susceptibility and so cannot give rise to dipolar (or pseudocontact) shifts, but can give rise to contact shifts. Because of slow electron spin relaxation in Gd^{3+}, NMR spectra of its complexes are usually extremely broad. Highly symmetrical complexes (T_d or O_h) of other ions also have isotropic magnetic susceptibility and so cannot give rise to dipolar shifts.

For an axially symmetrical complex (one with an axis of order 3 or higher, e.g. $Na_3[Ln(binol)_3]$) the pseudocontact shift for a nucleus is given by:

$$\Delta v = \frac{D\cos^2 \vartheta}{r^3}$$

where the constant D depends on $1/T^2$ and on the magnetic properties of the metal ion, and may be positive or negative depending on Ln. The distance r between the nucleus under observation and the Ln^{3+} ion, and the angle θ between the vector r and the principal symmetry axis of the complex, are defined in Figure 2.4, which also shows the variation of the term ($3\cos^2\theta-1$) with θ.

The magnitude and sign of Δv are highly sensitive to θ, as shown in Figure 2.4. When $\theta = 54.7°$ (half the tetrahedral angle) or $125.3°$, the term ($3\cos^2\theta - 1$) is equal to zero and so there is no pseudocontact shift. The magnitude and sign of the pseudocontact shift is clearly highly sensitive to geometry, depending on both r and θ and under favourable circumstances can give a great deal of structural information. The other important factor in determining the magnitude and sign of Δv is the magnetic properties of Ln^{3+}, and Figure 2.5 shows how the relative dipolar shift varies with Ln^{3+}.

- Most complexes of paramagnetic f-block ions display shifted NMR spectra.
- The shifting is almost entirely dipolar or 'pseudocontact' in origin.
- The sign and magnitude of Δv depends on geometrical factors as well as on the magnetic properties of the metal ion.
- Paramagnetic S state ions (e.g. Gd^{3+}) do not give rise to dipolar shifting.

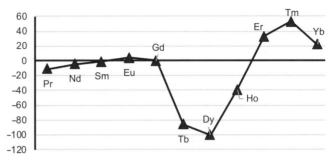

Figure 2.5 Variation in dipolar shift with Ln^{3+} (scaled to Dy = −100%)

2.5 Electronic absorption spectroscopy

Three types of electronic transition can occur for lanthanoid and actinoid systems. These are: f→f transitions, nf→(n+1)d transitions, and ligand→metal f charge transfer transitions. This discussion will be limited to lanthanoid complexes as these are much more straightforward than actinoid systems, where increased spin-orbit coupling and larger crystal field effects result in much more complex spectra.

The f→f transitions for lanthanoids and actinoids are, like d→d transitions of transition metal ions, electric-dipole forbidden by the Laporte ($\Delta L = \pm 1$) selection rule. Interaction with the ligand field or with vibrational states mixes in electronic states with different parity and so f→f transitions become possible. However, because of the small radial extent of the f-orbitals, these interactions are weak and the intensity of f→f transitions is therefore much lower than that of d→d transitions (a typical molar extinction coefficient ε for a f→f transition is 5 $M^{-1}cm^{-1}$), and the absorptions are very much sharper. The electronic absorption spectrum of aqueous $PrCl_3$, along with the appropriate transitions, is shown in Figure 2.6.

The electronic spectrum of aqueous Ce^{3+} shows no absorptions in the visible region of the spectrum: the $^2F_{5/2} \rightarrow {}^2F_{7/2}$ transition occurs at low energy in the infrared region of the spectrum, where it is masked by vibrational transitions, and there is an intense 4f→5d transition in the ultraviolet.

We have seen that there is very little interaction between f electrons and the crystal field, particularly for lanthanoid complexes, but electronic spectra do show evidence of some interaction with the crystal field in the form of a shifting to lower frequencies of the absorptions of complexes compared with those of the free ions. This has been explained in terms of a nephelauxetic effect resulting from a small degree (≤ 2.5% for lanthanoids) of metal-ligand covalent bonding.

Another crystal field effect is manifested in the form of hypersensitive transitions. The intensities of these transitions are found to be extremely sensitive to the ligand environment, varying by up to three orders of magnitude, depending

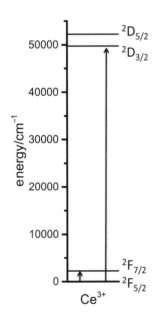

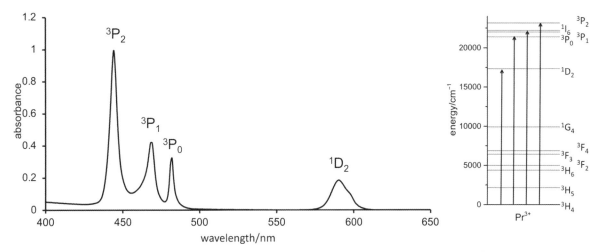

Figure 2.6 Electronic absorption spectrum of Pr^{3+} in aqueous solution

on the nature of the ligands. Not all f→f transitions are hypersensitive: of the lanthanoids Nd^{3+} and Er^{3+} show the greatest effects, and Ce^{3+}, Gd^{3+}, and Yb^{3+} show no hypersensitive transitions at all. The actinoids have much larger crystal field effects than the lanthanoids but Am^{3+} is the only actinoid for which hypersensitive transitions have been observed. There is no conclusive explanation of these hypersensitive transitions, but the most dramatic effects are seen for low symmetry complexes and those with polarizable ligands. Most hypersensitive transitions have $\Delta J = \pm 2$. Because of the dramatic intensity changes of hypersensitive transitions, they can be used to monitor complex formation in solution.

f→d transitions are Laporte allowed and therefore have much higher intensity than f→f transitions. They are also much broader, having a typical half-width of 1000 cm^{-1}. Ligand-to-metal charge transfer transitions are also Laporte allowed and therefore have high intensity. They are usually broader than f→d transitions; for easily reduced Ln^{3+} (Eu and Yb) the charge transfer transitions are at lower energy than the f→d transitions. For easily oxidized ligands they may tail into the visible region of the spectrum, giving rise to much more intensely coloured complexes. For example the tris(silylamides) of Eu and Yb (Chapter 4) are quite intense orange and yellow respectively, whereas other $[Ln\{N(SiMe_3)_2\}_3]$ are very pale in colour like their parent Ln^{3+} ions.

- The interaction of f-orbitals with the crystal field is small, especially for Ln complexes.
- f→f absorptions are sharp and have low extinction coefficients.
- Most f→f transitions are relatively insensitive to the nature of the ligands and the symmetry of the crystal field.
- Some 'hypersensitive' transitions (usually with $\Delta J \pm 2$) are very sensitive to the symmetry of the crystal field.

As predicted by the Landé interval rule, the separations between $^3P_2$ and $^3P_1$ and between $^3P_1$ and $^3P_0$ are in the ratio 2:1.

2.6 Lanthanoid ion luminescence

The term 'luminescence' is used to describe a whole range of phenomena which involve decay from an electronically excited state by emission of a photon. Fluorescence and phosphorescence are the emission of photons after a sample has been excited by electromagnetic radiation. The distinction between the two is that fluorescence is a spin-allowed process taking 10^{-6} to 10^{-12} s whereas phosphorescence involves a change in spin multiplicity and is a slower process, taking from 10^{-6} s to as much as several seconds. The term 'luminescence' will be used here to describe both fluorescence and phosphorescence.

Most lanthanoid ions luminesce in the solid state, and unlike luminescence from organic molecules, most lanthanoid emissions are sharp lines. This property has been used in lasers (e.g. the neodymium YAG laser) and in lanthanoid phosphors (e.g. Eu^{3+} and Tb^{3+} in fluorescent light phosphors). See Chapter 6 for applications of Ln luminescence.

Eu^{3+} and Tb^{3+} both display intense luminescence in the visible region (red for Eu^{3+} and green for Tb^{3+}) and this discussion will be limited to these two ions.

The appropriate energy level diagrams for Eu^{3+} and Tb^{3+} luminescence are shown in Figure 2.7. f→f transitions are Laporte forbidden and so excitation of Ln^{3+} to an emissive state by this route is not an efficient process. However, once the ion has been excited to the emissive state (the most important states are $^5D_0$ for Eu^{3+} and $^5D_4$ for Tb^{3+}), luminescence can occur providing other non-radiative processes do not take over.

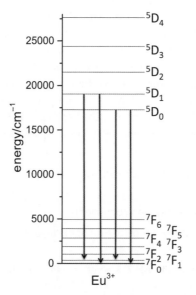
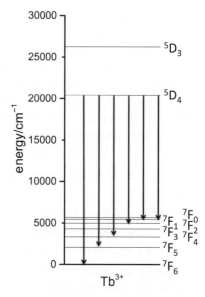

Figure 2.7 Eu^{3+} and Tb^{3+} luminescence

2.6.1 Vibrational de-excitation

The rate of non-radiative de-excitation is strongly related to the size of the energy gap between lowest emissive state and the highest state of the ground manifold; in the lanthanoid series this gap is largest for Eu^{3+} (12 150 cm^{-1}), Gd^{3+} (32 000 cm^{-1}) and Tb^{3+} (14 800 cm^{-1}). This energy gap can be bridged by weak vibronic coupling of the excited state with high frequency oscillators such as an overtone of the v(O–H) vibration of coordinated H_2O molecules. For Eu^{3+} the energy gap is bridged by three quanta of O–H vibrational energy as shown in Figure 2.8; four quanta are required to bridge the corresponding gap for Tb^{3+}. As a result of this facile de-excitation pathway, luminescence is not observed for aqueous solutions of Eu^{3+} or Tb^{3+}. Efficient luminescence requires that the lanthanoid ion is well separated from any high frequency oscillators.

There is a very significant isotope effect on the vibrational frequencies of H_2O vs. D_2O so that v(O–H) 3600 cm^{-1} whereas v(O–D) 2700 cm^{-1}. It therefore requires more quanta (approx. 5) of O–D vibration to bridge the gap than (O–H) (approx. 3), and vibrational quenching in H_2O is much more effective than in D_2O. Other high frequency oscillators (e.g. N–H) can also cause vibrational quenching.

2.6.2 Making use of the isotope effect in vibrational de-excitation

The difference between rates of vibrational de-excitation in D_2O and H_2O is the foundation of a technique for determining the number of coordinated H_2O molecules in a Eu^{3+} or Tb^{3+} complex in solution. This information often cannot be

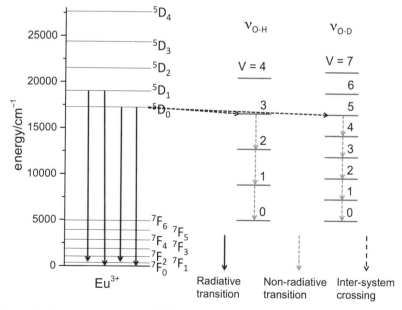

Figure 2.8 Vibrational quenching of Eu^{3+} luminescence via v(O–H) and v(O–D)

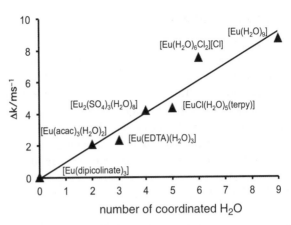

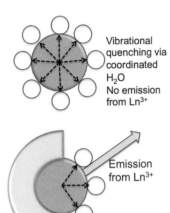

Vibrational quenching via coordinated H₂O

No emission from Ln³⁺

Emission from Ln³⁺

Figure 2.9 Plot of Δk vs. number of coordinated H₂O molecules for a series of Eu³⁺ complexes

obtained in any other way, and may be crucial to understanding processes such as catalysis or the action of MRI contrast agents (see Chapter 6).

Quenching via O–H vibrations is more effective than via O–D vibrations, and the magnitude of this difference depends linearly on the number of coordinated H_2O/D_2O ligands so that the largest effect is seen for $[Ln(H_2O)_9]^{3+}$ aqua ions.

Figure 2.9 plots Δk (k is 1/(excited state lifetime)) for a series of complexes that have known numbers of coordinated H_2O. When these complexes are dissolved in D_2O, the H_2O ligands exchange with D_2O. The excited state lifetimes are measured for each complex in H_2O and in D_2O and the difference Δk is plotted against the number of coordinated H_2O ligands. The maximum value of Δk is for $[Ln(H_2O)_9]^{3+}$ whereas for $[Eu(dipic)_3]$, which has no coordinated H_2O, Δk = 0. For an unknown complex, excited state lifetimes are measured in both H_2O and D_2O to give a value for Δk, and the number of H_2O ligands can be read off the graph.

2.6.3 Increasing the efficiency of lanthanoid luminescence

One way to increase the efficiency of luminescence is to reduce the number of high frequency oscillators (e.g. O–H, N–H) from the coordination sphere of the Ln ion, minimizing the possibility of vibrational de-excitation. The cryptand

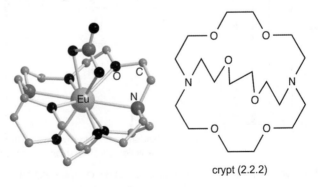

crypt (2.2.2)

Figure 2.10 Structure of $[Eu(2.2.2)(NO_3)]^{2+}$

ligand 2.2.2 encapsulates lanthanoid ions, reducing the number of coordinated H_2O molecules in aqueous solution to approximately two (Figure 2.10). As a consequence of this, the complexes $[Ln(2.2.2)]^{3+}$ (Ln = Eu or Tb) both luminesce in aqueous solution. The quantum yield where Ln = Tb is approximately an order of magnitude greater than that for the corresponding Eu complex, in part due to the higher energy of the emissive state for Tb^{3+}.

2.6.4 Sensitized luminescence

Another way to increase the efficiency of luminescence is to increase the efficiency of excitation. Because f→f transitions are Laporte forbidden, direct excitation to the emissive level cannot be an efficient process. An alternative method of excitation is via an organic ligand, usually a conjugated system, which has an excited triplet state higher in energy than the Ln^{3+} emissive state. On irradiation, the ligand molecule is excited into a singlet state, intersystem crossing then gives the ligand excited triplet state, and if the energy of this state is appropriate, fast intramolecular energy transfer to the lanthanoid emissive state can occur as shown in Figure 2.11.

This phenomenon was first observed for lanthanoid tris(β-diketonates) (Chapter 4), but much more spectacular results can be achieved using macrobicyclic polypyridine ligands such as bipy(2.2.2). This ligand has the correct electronic energy levels to act as an efficient antenna for excitation of Ln^{3+}, and due to its eight donor atoms it is able to prevent coordination of H_2O, thus avoiding vibrational de-excitation. A complex that is closely related to $[Eu\{bipy(2.2.2)\}]^{3+}$ is used in luminescent bioassays (see Chapter 6).

Figure 2.12 shows the excitation and emission spectra of a Tb^{3+}-containing metal organic framework compound. The excitation spectrum shows how the intensity of the most intense emission (in this case $^5D_4$ to $^7F_5$ at 545 nm) varies with the wavelength used for excitation.

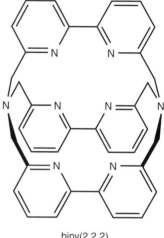

bipy(2.2.2)

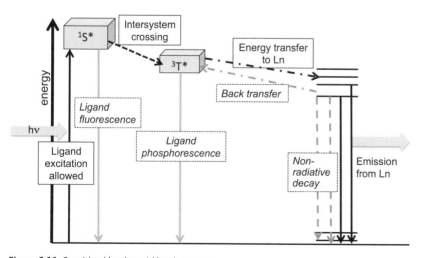

Figure 2.11 Sensitized lanthanoid luminescence

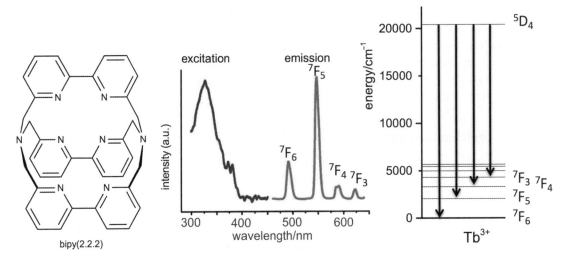

bipy(2.2.2)

Figure 2.12 Excitation and emission spectra for sensitized Tb^{3+} luminescence. Excitation spectrum obtained by monitoring emission at 545 nm ($^5D_4$ to $^7F_5$). Emission spectrum obtained by excitation at 327 nm. Reprinted with permission from J. Zhao, *Inorg. Chem.*, **55**, 73265–71. Copyright 2016 American Chemical Society.

2.7 **Summary**

The Russell–Saunders coupling scheme works reasonably well as a description of the coupling of angular momenta for Ln ions, and Hund's rules can be used to determine the ground state term symbols for these ions. Due to the very weak interactions between 4f orbitals and ligand orbitals, crystal field splittings are very small for Ln complexes, which means that Ln ions in complexes are very similar to the gas phase ions. The 'free ion' formula generally works well for calculating μ_{eff}, and electronic absorption spectra of Ln complexes are very similar to those of the gas phase ions. Some complexes of Ln^{3+} that have strongly axial crystal fields (e.g. $[LnPc_2]^-$) demonstrate single ion magnetism. Luminescence is observed for most Ln ions in the solid state, and with the correct choice of ligands, it is also observed in solution; Eu^{3+} and Tb^{3+} are the only two Ln^{3+} that luminesce in the visible region of the spectrum.

Due to much larger spin-orbit coupling, and the greater interaction of 5f orbitals with crystal fields than the corresponding 4f orbitals, electronic absorption spectra and magnetic properties of An ions are much more complicated than those of Ln ions.

2.8 **Exercises**

1. For each of the following ions, Nd^{3+}, Sm^{3+}, Eu^{3+}, and Gd^{3+}:
   (i) Use Hund's rules to derive the ground state term symbols (no cheating!)
   (ii) Use the 'free ion' formula to calculate the magnetic moment.

Table 2.2 gives magnetic data for a series of Ln β-diketonate complexes [LnL$_3$(H$_2$O)$_2$] (see Chapter 4). Compare the values of μ_{eff} that you calculated in (ii) with the values in Table 2.2.

2. A solution of Eu(OTf)$_3$ (OTf = CF$_3$SO$_3$) in MeCN was treated with ligand **L** (see Figure 2.13) to form a new complex **Z**.

The complexation of ligand **L** with Eu^{3+} was investigated by monitoring emission at 615 nm vs. mole fraction of Eu^{3+}. A Job's plot of emission intensity at 615 nm vs. mole fraction Eu^{3+} is shown in Figure 2.13. Use this plot to determine the stoichiometry of the complex formed between Eu^{3+} and **L**.

Explain the mechanism by which **Z** displays luminescence. Why is there no measurable emission when mole fraction Eu^{3+} = 1?

The excited state lifetimes for **Z** were measured in both H$_2$O and D$_2$O and were found to be the same in both solvents. What can be deduced from this observation?

See *J. Am. Chem. Soc.*, 2009, **131**(28): 9636–7.

HL

Table 2.2 μ_{eff} for [LnL$_3$(H$_2$O)$_2$]

Ln	μ_{eff}/μ_B (298K)	μ_{eff}/μ_B (99K)
Nd	3.39	3.06
Sm	1.67	1.04
Eu	3.26	2.18
Gd	8.03	7.85

Data from *Transition Met. Chem. 1983*, **8**, p. 298.

Emission intensity at 615 nm vs. mole fraction Eu^{3+}

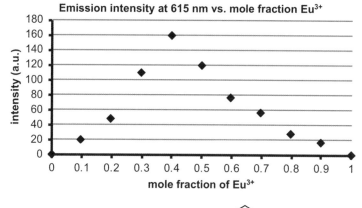

Figure 2.13 Job's plot for complexation of ligand L with Eu^{3+}

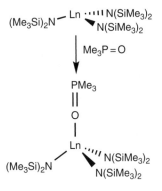

3. Lanthanoid tris(silylamides) $[Ln\{N(SiMe_3)_2\}_3]$ (see Chapter 4) adopt a trigonal planar structure in solution. On addition of one equivalent of $Me_3P=O$, a trigonal pyramidal complex is formed.

^1H NMR data (298K) for $[Ln\{N(SiMe_3)_2\}_3]$ and $[Ln\{N(SiMe_3)_2\}_3(Me_3P=O)]$ are given in Table 2.3. Explain the different chemical shift values for Ln = La, Pr, and Eu, and for $[Ln\{N(SiMe_3)_2\}_3]$ vs. $[Ln\{N(SiMe_3)_2\}_3(Me_3P=O)]$.

Table 2.3 ^1H NMR data for $[Ln\{N(SiMe_3)_2\}_3]$ and $[Ln\{N(SiMe_3)_2\}_3(Me_3P=O)]$

Ln	$[Ln\{N(SiMe_3)_2\}_3]$	$[Ln\{N(SiMe_3)_2\}_3(Me_3P=O)]$	
		$SiMe_3$	$Me_3P=O$
La	0.25	0.53	0.85, d
Pr	−8.64	0.09	−22.81, d
Eu	6.43	-0.45	23, d

Data from *Polyhedron, 1982,* **1**, p. 307.

2.9 **Further reading**

1. Ishikawa, N., et al., 'Lanthanide double-decker complexes functioning as magnets at the single-molecular level'. *Journal of the American Chemical Society*, 2003,**125**(29): 8694–5.
2. Woodruff, D.N., R.E.P. Winpenny, and R.A. Layfield, 'Lanthanide single-molecule magnets'. *Chemical Reviews*, 2013, **113**(7): 5110–48.
3. Bunzli, J.C.G., 'On the design of highly luminescent lanthanide complexes'. *Coordination Chemistry Reviews*, 2015, **293**: 19–47.
4. McAdams, S.G., et al., 'Molecular single-ion magnets based on lanthanides and actinides: Design considerations and new advances in the context of quantum technologies'. *Coordination Chemistry Reviews*, 2017, **346**: 216–39.

3 Binary compounds: oxides and halides

3.1 Introduction

Lanthanoid and actinoid oxides and halides are important starting materials for many industrial and laboratory scale processes, e.g. for production of the elements (Chapter 7) and for synthesis of many coordination and organometallic complexes (Chapters 4 & 5). They also have many applications in their own right: e.g. lanthanoid oxides as catalysts or catalyst supports, and UO_2 as a nuclear fuel (Chapter 6). UF_6 is the key compound in isotopic enrichment of uranium for use in nuclear reactors (Chapters 6 & 7), and as such is probably the most studied actinoid halide. This chapter will cover a selection of compounds to illustrate the most important features, and we will see how the structures and stoichiometries of oxides and halides reflect trends in ionic radii and oxidation states.

3.2 Lanthanoid oxides

All of the lanthanoid metals react readily with oxygen to form oxides, which are the most stable compounds of these elements (e.g. $\Delta_fH°$ for $La_2O_3 = -1794$ kJ mol^{-1}). Oxides of Ln in +2, +3, and +4 oxidation states are known.

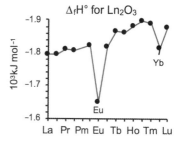

3.2.1 Lanthanoid sesquioxides Ln_2O_3

The sesquioxides Ln_2O_3 are known for all of the lanthanoids, and for most Ln these are the most stable oxides (Ce, Pr, and Tb are exceptions as they have accessible +4 oxidation states). In the case of Ce, it is difficult to prepare stoichiometric Ce_2O_3, and anaerobic conditions must be used. A non-stoichiometric oxide $CeO_{1.50-1.53}$ is also known; it differs from stoichiometric Ce_2O_3 both in structure and in the fact that it can be pyrophoric.

The influence of Ln^{3+} radius on structures is summarized in Figure 3.1.

There are three structural types for Ln_2O_3, and the structure adopted depends on Ln^{3+} radius: the A-type structure adopted by early Ln has 7-coordinate Ln^{3+} whereas the C-type structure adopted by later Ln has 6-coordinate Ln^{3+}. The reactivity depends on the structure adopted: all Ln_2O_3 react with atmospheric CO_2

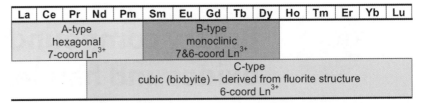

La	Ce	Pr	Nd	Pm	Sm	Eu	Gd	Tb	Dy	Ho	Tm	Er	Yb	Lu
A-type hexagonal 7-coord Ln^{3+}					B-type monoclinic 7&6-coord Ln^{3+}									
				C-type cubic (bixbyite) – derived from fluorite structure 6-coord Ln^{3+}										

Figure 3.1 Structures of Ln_2O_3

and H_2O to form carbonates, hydrated carbonates, and hydroxides, but these reactions are much more facile for A-type than for C-type Ln_2O_3.

Ln_2O_3 are used as catalysts and catalyst supports, particularly in the petrochemical industry. Rare earth sesquioxides have dielectric constants κ in the range 12.5 (cubic Lu_2O_3) to 17 (hexagonal La_2O_3). The values of κ are closely related to the frequency of the dominant IR lattice mode, and these relatively high values, combined with their thermodynamic stability, have led to Ln_2O_3 being serious candidates for replacing SiO_2 as the gate dielectric in MOSFETs (Metal Oxide Semiconductor Field Effect Transistors).

Metallic Sm, Eu and Yb are manufactured by reduction of Ln_2O_3.

3.2.2 Higher oxides of the lanthanoids LnO_{2-x}

We have seen in Chapter 1 that Ce has the most accessible +4 oxidation state of all the lanthanoids, and CeO_2 (also known as 'ceria') is the most stable oxide for Ce. Like many other metal dioxides, CeO_2 adopts the fluorite structure with 8-coordinate Ce.

The +4 oxidation state is also accessible for Pr and Tb, and these elements display a very rich chemistry with O. The most common oxides for Pr and Tb are $Pr_{12}O_{22}$ and Tb_4O_7, which contain both Ln^{3+} and Ln^{4+}. These compounds are very intensely coloured due to charge-transfer transitions. Numerous other phases with stoichiometries between $Ln_{1.0}O_{1.5}$ and $Ln_{1.0}O_{2.0}$ can be obtained at high temperatures and O_2 pressures. All of these 'higher oxides' adopt fluorite-related structures.

3.2.3 Unique properties of CeO_2

Two factors contribute to the unique properties of CeO_2: firstly, the fluorite structure is extremely stable over quite a wide range of compositions, and secondly, the +3 and +4 oxidation states are both accessible for Ce. Oxygen vacancies can be created in the lattice by reduction of Ce^{4+} to Ce^{3+} accompanied by release of O_2 as shown below:

$$2Ce_{Ce} + O_o \rightarrow V_o^{\cdot\cdot} + 2Ce'_{Ce} + \tfrac{1}{2}O_2$$

A particularly stable composition is $CeO_{1.714}$ (which is equivalent to Ce_7O_{12}). In this composition, one oxygen vacancy is surrounded by an octahedral array of six $[Ce_{0.5}O]$ units, while retaining the fluorite structure as shown in Figure 3.2.

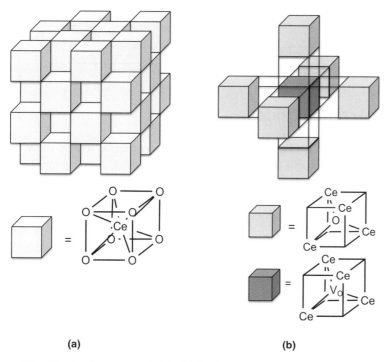

Figure 3.2 (a) The fluorite structure of CeO_2 (b) The O-vacancy structure in $CeO_{1.714}$

As the number of oxygen vacancies increases to give the composition Ce_2O_3, the cubic C-type structure, also closely related to fluorite, is obtained. CeO_2 can be regenerated under oxidizing conditions, and this reversible uptake and release of O_2 is key to catalytic applications of ceria, some of which will be discussed in Chapter 6.

3.2.4 Lanthanoid monoxides LnO

Lanthanoid monoxides LnO (Ln = La, Ce, Pr, Nd, and Sm) can be prepared by reduction of Ln_2O_3 with metallic Ln at high pressure, but apart from EuO these are rather unstable and readily disproportionate. However, despite this instability, several LnO have been structurally characterized, and they all adopt the rocksalt structure with 6-coordinate Ln ions. A selection of Ln–O distances are shown in Table 3.1, and these distances suggest that CeO and PrO are very different from EuO and YbO. In Chapter 1 we saw that ionic radii for the lanthanoid series decrease from La to Lu, but Ln–O distances are *shorter* in CeO and PrO than in EuO. The explanation of this apparent anomaly is that, despite the stoichiometry, only EuO and YbO contain Ln^{2+} ions; other LnO contain Ln^{3+} and are best formulated as $[Ln^{3+}][e^-][O^{2-}]$ with delocalized electrons in a 5d band. Consistent with this formulation, EuO and YbO are insulators whereas CeO and PrO crystallize with a metallic lustre and are conductors. Similar effects are seen for Ln dihalides (see Section 3.4).

Table 3.1 Ln–O distances in lanthanoid monoxides

	Ce	Pr	Eu	Yb
Ln–O distance/Å	2.545	2.516	2.571	2.439

3.3 Actinoid oxides

As we have seen in Chapter 1, the actinoid elements display a wider range of oxidation states than the lanthanoids. This results in a more diverse array of oxides, as summarized in Table 3.2. Oxides with actinoid oxidation states from +3 to +6 are known, and there are also many examples of non-stoichiometric oxides. UO_2 and PuO_2 have been particularly well studied due to their use in nuclear fuels (see Chapter 6).

The relatively simple summary presented in Table 3.2 hides a multitude of complexity. For example there are up to a dozen known phases of uranium oxides with compositions between UO_2 and UO_3, many of which are formed by random incorporation of additional O into the fluorite lattice of UO_2. A common feature of several higher oxides of U, Np, and Pu is the presence of actinyl AnO_2^{2+} moieties with short An–O distances that are consistent with covalent bonding (Figure 3.3). (See Chapter 4 for more about AnO_2^{2+} ions.)

Table 3.2 Summary of actinoid oxides

Ac	Th	Pa	U	Np	Pu	Am	Cm	Bk	Cf	Es
Ac_2O_3					Pu_2O_3*	Am_2O_3	Cm_2O_3	Bk_2O_3	Cf_2O_3	Es_2O_3
	ThO_2	PaO_2*	UO_2	NpO_2	PuO_2	AmO_2*	CmO_2*	BkO_2	CfO_2*	
		Pa_2O_5		Np_2O_5						
			UO_3							
			U_3O_8							

* = non-stoichiometric

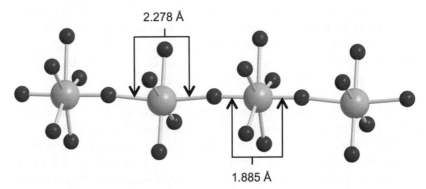

Figure 3.3 Chain of alternating UO_6 octahedra and UO_7 pentagonal bipyramids in β-U_3O_8. Note the short axial U–O distances in the pentagonal bipyramids.

3.4 Lanthanoid halides

Anhydrous lanthanoid halides are industrially important as starting materials for the manufacture of metallic Ln by reduction (either electrolytic or chemical). Anhydrous $LnCl_3$ and LnI_2 are important starting materials for the synthesis of coordination and organometallic compounds of the lanthanoids (Chapters 4 & 5). The fluorides LnF_3 are of interest for optical applications (e.g. as phosphors, and for frequency doubling and quantum cutting). The bonding in lanthanoid halides is essentially ionic and so the radii of both Ln and halide ions determine the structural trends. The trihalides LnX_3 are the best known, but as expected, fluoride stabilizes high oxidation states (e.g. in CeF_4), and iodide stabilizes lower oxidation states (e.g. in SmI_2 and EuI_2).

Metallic Ce was first prepared by Mosander 1839 by K reduction of molten $CeCl_3$ under an atmosphere of H_2.

3.4.1 Lanthanoid(III) halides

Lanthanoid trihalides LnX_3 are known for all Ln and for X = F, Cl, Br, and I. Figure 3.4 shows $\Delta_fH°$ for these compounds. As expected the magnitude of $\Delta_fH°$ for all of the trihalides is at a minimum for Eu and Yb, which have the most stable +2 oxidation states due to the half-filled and filled 4f sub-shells for Eu^{2+} and Yb^{2+} respectively. The importance of the lattice enthalpy contributions is reflected in the relative magnitudes of $\Delta_fH°$ for fluorides, chlorides, bromides, and iodides. The anhydrous chlorides, bromides, and iodides are all highly soluble in water and are very hygroscopic, in part due to the large enthalpies of hydration for Ln^{3+} (e.g. $\Delta_{hyd}H°$ for Gd^{3+} -3517 kJ mol^{-1}). However, LnF_3 are insoluble in water and

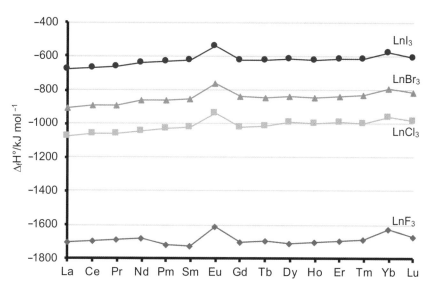

Figure 3.4 Standard enthalpies of formation for lanthanoid trihalides. (Data for LnF_3 from Kim, K.-Y. and Johnson, C.E. *Journal of Chemical Thermodynamics*, 1981, **13**: 13–25. Other data from Cordfunke, E.H.P. and Konings, R.J.M. *Thermochimica Acta*, 2001, **375**: 17–50.)

non-hygroscopic due to the large $\Delta_L H°$ for these compounds. Anhydrous LnF_3 are used as starting materials for the manufacture of metallic Ln by reduction with Ca.

The structures of LnX_3 are determined by ionic radii of Ln^{3+} and X^- and are summarized in Figure 3.5.

Figure 3.6 shows the coordination geometry around 11-coordinate La^{3+} in the tysonite structure and 9-coordinate Y^{3+} in the YF_3 structure. The tri-capped trigonal prism is common throughout f-element chemistry (e.g. $[Ln(H_2O)_9]^{3+}$ described in Chapter 4).

3.4.2 **Lanthanoid (IV) halides**

As expected, the highest oxidation states are stabilized by fluoride, and LnF_4 are known for Ce, Pr, and Tb, the lanthanoids with the most accessible +4 oxidation states. They can be prepared by oxidation of the metal (in the case of Ce) or LnF_3 with F_2. All of the LnF_4 adopt the 'ZrF_4' structure, with 8-coordinate Ln^{4+} in a distorted square antiprismatic coordination geometry. The square antiprism is a common geometry throughout f-element coordination chemistry (e.g. $[Ln(H_2O)_8]^{3+}$ described in Chapter 4).

Cl^- is much more susceptible to oxidation than F^- and so simple binary $LnCl_4$ compounds are unknown, even for Ce. However, $[CeCl_6]^{2-}$ can be stabilized in the presence of large counterions such as $[Et_4N]^+$, and salts of this type can be used as starting materials for synthesis of Ce(IV) coordination compounds.

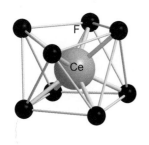

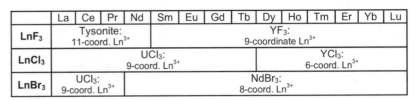

	La	Ce	Pr	Nd	Sm	Eu	Gd	Tb	Dy	Ho	Tm	Er	Yb	Lu
LnF₃	Tysonite: 11-coord. Ln³⁺							YF₃: 9-coordinate Ln³⁺						
LnCl₃	UCl₃: 9-coord. Ln³⁺									YCl₃: 6-coord. Ln³⁺				
LnBr₃	UCl₃: 9-coord. Ln³⁺			NdBr₃: 8-coord. Ln³⁺										

Figure 3.5 Structures of lanthanoid trihalides

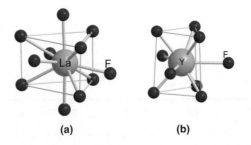

(a)   (b)

Figure 3.6 Coordination of Ln^{3+} in LnF_3 (a) fully-capped trigonal prism in LaF_3 (tysonite) structure (b) tri-capped trigonal prism in YF_3 structure

3.4.3 **Lanthanoid(II) halides**

Ionic $LnCl_2$ can be prepared by reduction of $LnCl_3$ with Ln, and are known for Nd, Sm, Eu, Dy, Tm, and Yb. The relative stabilities of $LnCl_2$ with respect to $LnCl_3$ can be understood by considering the thermochemical cycle shown in Figure 3.7(a). The lattice enthalpy terms for both $LnCl_2$ and $LnCl_3$ vary relatively smoothly with ionic radius along the series, so the most important terms in the cycle are the third ionization potential and the heat of atomization for Ln. The value of 3rd IP $- \Delta_{at}H°$, plotted in Figure 3.7(b), gives an indication of the relative stability of $LnCl_2$ compared with $LnCl_3$, and is at a maximum for Eu and Yb, consistent with the stability of $EuCl_2$ and $YbCl_2$. The stability of $LnCl_2$ is at a minimum for La, Gd, and Lu.

$EuCl_2$ and $YbCl_2$ were the first compounds of Ln^{2+} to be discovered.

The sesquichloride Gd_2Cl_3 can be formulated as $[Gd^{3+}]_2[e^-]_3[Cl^-]_3$ with 1.5 electrons per Gd^{3+} involved in metal–metal interactions. The structure is made up of infinite chains of edge-linked Gd_6 octahedra aligned parallel to the b axis. Each Gd is coordinated to four other Gd atoms (the nearest at 3.5 Å, the others at between 3.7 and 3.8 Å) and to five Cl atoms at between 2.72 and 2.88 Å. Figure 3.8 shows two views of the Gd_2Cl_3 structure.

Ionic dibromides and diiodides are known for Nd, Sm, Eu, Dy, Tm, and Yb. The diiodides of Sm, Eu, and Yb are particularly stable and they are useful starting materials for organometallic and coordination chemistry of these elements in their +2 oxidation states. SmI_2 is a useful 1-electron reducing agent for organic chemistry (see Chapter 6). The diiodides of La, Ce, Pr, and Gd are very different: they have metallic properties and are best formulated as $[Ln^{3+}]$ $[I^-]_2[e^-]$ with the extra electrons accommodated in delocalized bands formed from 5d orbitals. We have already seen the occupation of 5d rather than 4f orbitals in lanthanoid monoxides, and there will be further examples in organometallic and coordination complexes of Ln(II) for La, Ce, Pr, Gd (e.g. $[Cp_3Ln]^-$, Chapter 5).

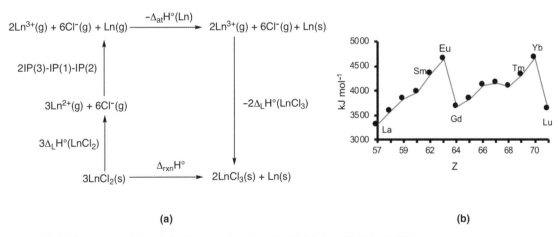

Figure 3.7 (a) Thermochemical cycle for disproportionation of $LnCl_2$ (b) plot of 3rd IP $- \Delta_{at}H°$ for Ln

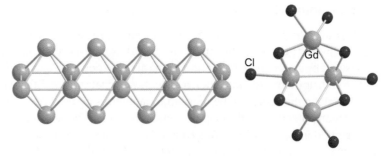

Figure 3.8 Structure of Gd_2Cl_3. Left: edge-sharing octahedra of Gd^{3+}; right: view along Gd_2Cl_3 chain.

3.5 Actinoid halides

Actinoid halides are technologically important in the production of metallic An, and UF_6 is particularly important for isotopic separation of $^{235/238}U$ (Chapter 7). As in lanthanoid chemistry, anhydrous actinoid halides are key starting materials for synthesis of coordination and organometallic compounds. The range of actinoid halides reflects the available oxidation states from An^{2+} to An^{6+}, and their structures reflect trends in An^{n+} radii.

3.5.1 Uranium fluorides

Table 3.3 summarizes properties of uranium fluorides in oxidation states from +3 to +6. The coordination number of U decreases with decreasing radius of U^{n+} and the covalent character of the U–F bonds increases from UF_3 (a typical ionic solid) to UF_6 (a typical molecular crystal).

The crystal structure of UF_6 shows it to consist of isolated octahedral UF_6 molecules, consistent with its high volatility, and calculations indicate that there is a significant degree of covalency in the bonding. In an octahedral ligand field, uranium has 5f, 6d, 6p, and 7p orbitals with the correct symmetry to take part in both σ- and π-bonding with F.

UF_6 is the most thoroughly studied halide of uranium because of its importance in isotope separation (see Chapter 7).

UF_6 packing

All the AnF_6 are low melting volatile solids with bps of 56.54, 55.18, and 62.26 °C for UF_6, NpF_6, and PuF_6 respectively.

Table 3.3 Properties of uranium fluorides

	UF_3	UF_4	UF_5	UF_6
c.n. of U	11	8	7 or 6	6
c.n. of F	3, 4	2	1, 2	1
$\Delta_fH°$/kJ mol^{-1}	−1470	−1884	−2056	−2190
mp/K	1700 (dec)	1309	621	337

3.6 **Summary**

Binary oxides and halides of the f-elements are important as starting materials for manufacture of the elements, as starting materials for synthesis of other compounds and complexes, and as materials in their own right. For example, CeO_2 has a very rich chemistry due to the accessibility of both +3 and +4 oxidation states for Ce, and this results in practical applications (see Chapter 6). The oxide chemistry for the actinoid elements from Pa to Pu is also very rich due to the accessibility of a range of oxidation states. Structures and stoichiometries of oxides and halides reflect trends in ionic radii and in relative stabilities of oxidation states.

3.7 **Exercises**

1. $Pr_{12}O_{22}$ is a mixed valence oxide containing Pr^{4+} and Pr^{3+}. Calculate the value of x in $(Pr^{4+})_x(Pr^{3+})_{12-x}O_{22}$. When $Pr_{12}O_{22}$ is added to aqueous hydrochloric acid, the black solid dissolves with effervescence to form a pale green solution. What is in the solution, and what is the gas that is evolved?

2. $\Delta_fH°$ for all Ln_2O_3 are large and negative, but the magnitudes of $\Delta_fH°$ for Ln = Eu or Yb are significantly smaller than for other Ln. Why is this?

3. The structure of Np_2O_5 consists of chains of vertex-linked alternating NpO_7 pentagonal bipyramids and NpO_6 octahedra that are joined together to form sheets. Part of a chain is shown in Figure 3.9.

    What is the significance of the distances shown in Table 3.4? (Hint: see Section 3.3.)

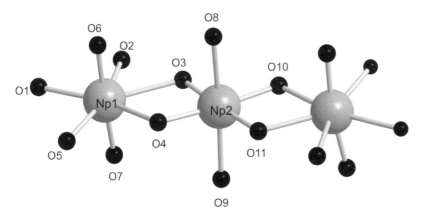

Figure 3.9 Chain of alternating pentagonal bipyramids and octahedra in the Np_2O_5 structure

Table 3.4 Np–O distances in Np_2O_5

distance/Å		distance/Å	
Np1–O1	2.350	Np2–O3	2.130
Np1–O2	2.596	Np2–O4	1.959
Np1–O3	2.441	Np2–O8	2.349
Np1–O4	2.596	Np2–O9	2.305
Np1–O5	2.350	Np2–O10	1.966
Np1–O6	1.885	Np2–O11	2.156
Np1–O7	1.885		

3.8 Further reading

1. Adachi, G.-Y. and N. Imanaka, 'The binary rare earth oxides'. *Chemical Reviews*, 1998, **98**(4): 1479–514.
2. Meyer, G., 'Reduced halides of the rare-earth elements'. *Chemical Reviews*, 1988, **88**(1): 93–107.
3. Meyer, G., 'Small cause – Great effect: What the $4f^{n+1} 5d^0 \rightarrow 4f^n 5d^1$ configuration crossover does to the chemistry of divalent rare-earth halides and coordination compounds'. *Journal of Solid State Chemistry*, 2019, **270**: 324–34.

4 Coordination chemistry

4.1 Introduction

This chapter will cover complexes of f-elements excluding those with Ln- or An-carbon bonds. Both lanthanoid and actinoid ions generally behave as hard Lewis acids and so predominantly form complexes with hard O- and N-donor ligands. However there will be some examples with soft donor ligands such as S.

We have already seen that the ionic radii of both lanthanoids and actinoids are larger than those of most d-transition metals, and this results in generally higher coordination numbers (8, 9, and 10 are common). The lanthanoid series presents a unique opportunity to investigate the effect of metal ion radius on coordination chemistry and there are many examples where coordination numbers change along the series. The coordination geometries are dictated by steric factors and as a consequence are often rather irregular.

For the lanthanoids we will see that the majority of the complexes have Ln in the +3 oxidation state, though there are many examples of Eu^{2+} and Yb^{2+}, and a growing number of complexes of other Ln in the +2 oxidation state. Ce is the only lanthanoid for which the +4 oxidation state is accessible in solution. The 4f orbitals are essentially core orbitals and do not take part in Ln-ligand bonding, so crystal field effects are negligible for the lanthanoids, in contrast to d-transition metal chemistry. The essentially ionic nature of Ln-ligand bonding results in irregular coordination geometries that are dictated only by ligand steric effects.

The actinoids present more variety than the lanthanoids in their coordination chemistry, particularly the early actinoids Pa to Am. We have seen in Chapter 1 that these elements have several accessible oxidation states, and that there is a possibility of 5f and 6d orbital participation in bonding, which can result in appreciable covalent contributions to An-ligand bonding. Beyond Am, the +3 oxidation state becomes increasingly important, and 5f orbital contribution to bonding becomes insignificant so that in most respects the coordination chemistry of the late actinoids resembles that of their lanthanoid counterparts.

This chapter will start with a discussion of coordination chemistry in aqueous solution, and then move on to the more varied coordination chemistry that is seen in non-aqueous solution. We will see the importance of chelating and macrocyclic ligands in f-element coordination chemistry, particularly in aqueous solution.

4.2 Lanthanoid and actinoid ions in aqueous solution

4.2.1 Structures of lanthanoid aqua ions

For all of the lanthanoids, the +3 oxidation state is the most stable oxidation state in aqueous solution. Apart from Eu^{2+}, all other Ln^{2+} are oxidized by water. Ce^{4+} is the only Ln^{4+} that is accessible in aqueous solution; it is a very powerful oxidizing agent ($E° = 1.3$ V for Ce^{4+}/Ce^{3+} in 1M HCl) and so should oxidize water, but it is stable for kinetic reasons.

Ln^{3+} form the aqua ions $[Ln(H_2O)_n]^{3+}$ in aqueous solution. For La^{3+} to Nd^{3+}, n =9, and n = 8 for Gd^{3+} to Lu^{3+}. For Pm^{3+} to Eu^{3+} there is an equilibrium between $[Ln(H_2O)_8]^{3+}$ and $[Ln(H_2O)_9]^{3+}$. The change in coordination number of aqua ions around the middle of the lanthanoid series has important effects on the coordination chemistry in aqueous solution. $[Ln(H_2O)_n]^{3+}$ are extremely labile in solution: for Gd^{3+}(aq) the H_2O exchange rate $k \approx 10^8$ s^{-1}.

In the solid state $[Ln(H_2O)_9]^{3+}$ is known for all of the lanthanoids, but $[Ln(H_2O)_8]^{3+}$ is sometimes observed for the smaller late lanthanoids. The structures of $[La(H_2O)_9]^{3+}$ and $[Yb(H_2O)_8]^{3+}$ (from a structural study of lanthanoid triflates) are shown in Figures 4.1 and 4.2.

$[La(H_2O)_9]^{3+}$ adopts a tri-capped trigonal prismatic coordination geometry, whereas $[Yb(H_2O)_8]^{3+}$ is square-antiprismatic. EXAFS studies have confirmed that the structures of aqua ions in aqueous solution are analogous to those in the solid state. The decrease in coordination number as the Ln^{3+} radius decreases from

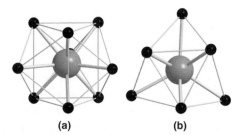

(a)   (b)

Figure 4.1 Tri-capped trigonal prism structure of $[La(H_2O)_9]^{3+}$: (a) perpendicular to 3-fold axis; (b) along 3-fold axis

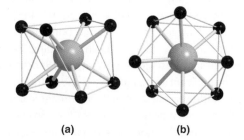

(a)   (b)

Figure 4.2 Square antiprism structure of $[Yb(H_2O)_8]^{3+}$: (a) perpendicular to 4-fold axis; (b) along 4-fold axis

La^{3+} to Yb^{3+} is a common phenomenon in lanthanoid coordination chemistry, and an illustration of one of the consequences of the lanthanoid contraction. There is also a decrease in effective ionic radius with decreasing coordination number (for La^{3+} 9-coordinate radius = 135.6 pm, 8-coordinate radius = 130.0 pm; for Yb^{3+} 9-coordinate radius = 118.2 pm, 8-coordinate radius = 112.5 pm). The combination of lanthanoid contraction and decreased coordination number results in a difference in Ln–O distances of 2.29 pm between $[La(H_2O)_9]^{3+}$ and $[Yb(H_2O)_8]^{3+}$. The shorter and therefore stronger Ln–O bonds in $[Yb(H_2O)_8]^{3+}$ compared with $[La(H_2O)_9]^{3+}$ have important consequences for the aqueous solution chemistry of these two species.

4.2.2 Brønsted acidity of lanthanoid aqua ions

Lanthanoid ions are hard Lewis acids and in aqueous solution they display Brønsted acidity as shown in Figure 4.3, which increases with increasing charge and decreasing ionic radius.

Lanthanoid hydroxides $Ln(OH)_3$ are very insoluble in aqueous solution, and solubility decreases from La to Lu with decreasing Ln^{3+} radius: K_{sp} for $La(OH)_3 \approx 4 \times 10^{-19}$, corresponding to a solubility of ca. 5 mg per litre; K_{sp} for $Lu(OH)_3 \approx 2.4 \times 10^{-24}$, corresponding to a solubility of ca. 0.3 mg per litre. The insolubility of $Ln(OH)_3$ is a major challenge for lanthanoid coordination chemistry in aqueous solution: except in the presence of very strongly coordinating ligands, $Ln(OH)_3$ precipitates out of solution at pH > 7.5 for La or pH > 5.5 for Lu as shown in Figure 4.4. The increased charge and decreased radius of Ce^{4+} (6-coordinate

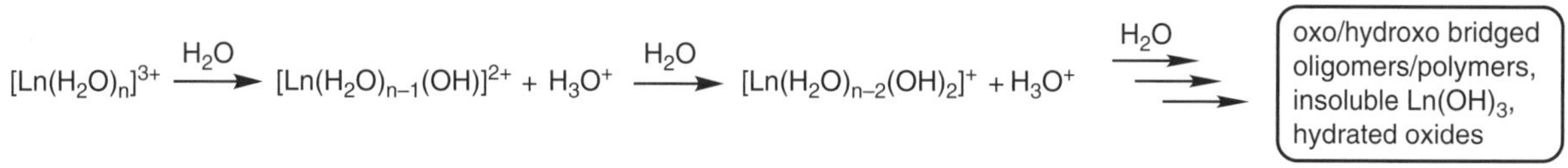

$$[Ln(H_2O)_n]^{3+} \xrightarrow{H_2O} [Ln(H_2O)_{n-1}(OH)]^{2+} + H_3O^+ \xrightarrow{H_2O} [Ln(H_2O)_{n-2}(OH)_2]^+ + H_3O^+ \xrightarrow{H_2O} \boxed{\text{oxo/hydroxo bridged oligomers/polymers, insoluble } Ln(OH)_3, \text{ hydrated oxides}}$$

Figure 4.3 Brønsted acidity of $[Ln(H_2O)_n]^{3+}$ in aqueous solution

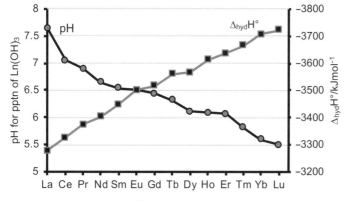

Figure 4.4 Hydration enthalpies for Ln^{3+} and pH at which $Ln(OH)_3$ precipitates from aqueous solution

radius = 101 pm) results in increased Brønsted acidity, and $Ce(OH)_4$ precipitates out of aqueous solution at pH > 2.6.

4.2.3 Actinoid aqua ions

An^{3+} aqua ions are very similar to their Ln^{3+} analogues, but the variety of oxidation states available to the early actinoids results in more varied aqua ion chemistry than for the lanthanoids.

The +4 oxidation state is the only stable state for Th, and is one of several accessible states for U, Np, and Pu. The increased charge and decreased radius for An^{4+} compared with An^{3+} means that $[An(H_2O)_n]^{4+}$ (n = 10–11) are much more Brønsted acidic than their An^{3+} analogues so that appreciable hydrolysis occurs at pH > 2, ultimately forming highly insoluble hydrated oxides. The rate of hydrolysis is $Th^{4+} < U^{4+} < Np^{4+} < Pu^{4+}$ with increasing value of $1/r$.

An^{5+} is important for Pa, Np, Pu, and Am, and An^{6+} is important for U and Pu. The very high charge/radius ratio for these ions means that their simple aqua ions are unknown: hydrolysis occurs to form actinyl ions $[AnO_2(H_2O)_n]^{x+}$ (x = 1 or 2) as shown in Figure 4.5. Pa^{5+} is an exception: it has a unique tendency to hydrolyse and at pH > 1 it forms polymeric oxohydroxides.

The rates of hydrolysis of An ions in aqueous solution decrease in the order:

$$An^{4+} > \left[AnO_2\right]^{2+} > An^{3+} > \left[AnO_2\right]^{+}$$

The effective charge on the $An^{6+/5+}$ ions in $[AnO_2]^{2+/1+}$ are reduced to 3.3 and 2.2 respectively due to the two O^{2-} that are bonded to each An. The bonding in actinyl ions $[AnO_2]^{x+}$ is discussed more fully below.

4.2.4 Actinyl ions $[AnO_2]^{x+}$ (x = 1 or 2)

The very high charge:radius ratio of $An^{5+/6+}$ ions means that $[An(H_2O)_n]^{5+/6+}$ are far too strongly Brønsted acidic to exist in aqueous solution, and they are immediately hydrolysed to form the actinyl ions $[AnO_2]^{1+/2+}$ (An = U, Np, Pu, and Am). The actinyl ions are remarkably persistent species with very strong An–O bonds, and they have been the subjects of numerous structural and theoretical studies. As well as having an extensive coordination chemistry, actinyl ions also exist in some solid state oxide structures (see Chapter 3).

In $[UO_2(H_2O)_5]^{2+}$ the average $U–O_{axial}$ distance is 175 pm (cf average $U–O_{equatorial}$ distance of 241 pm) and the $O_{axial}–U–O_{axial}$ angle (178.5°) is essentially linear. The short $U–O_{axial}$ bonds are inert whereas the $U–O_{equatorial}$ bonds are very labile; this results in an extensive coordination chemistry of $[UO_2]^{2+}$ in which the uranyl ion persists as a discrete entity throughout.

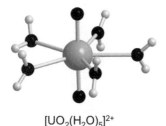

$[UO_2(H_2O)_5]^{2+}$

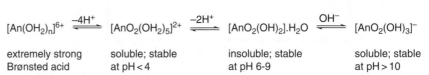

Figure 4.5 Hydrolysis of $[An(H_2O)_n]^{6+}$ in aqueous solution

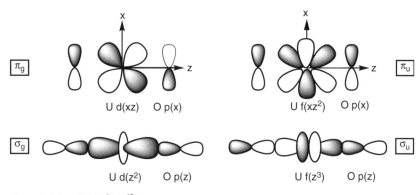

Figure 4.6 Bonding in $[UO_2]^{2+}$

The unique properties of axial An–O bonds in the actinyl ions are a consequence of covalent bonding in early An species.

The principal valence orbitals of U are 6d and 5f, and there is good spectroscopic evidence for 6p-5f(z^3) hybridization. The U 6d(xz), 5f(xz^2) and 5f(yz^2) orbitals have the correct symmetry to form π-bonding interactions with O 2p(x) and 2p(y) orbitals, and σ-bonding interactions can form between O 2p(z) orbitals and U 6d(z^2) and 5f(z^3) orbitals, resulting in formal U≡O triple bonds. These interactions are shown in Figure 4.6.

The uranyl-type structure also occurs in the solid state in higher oxides of uranium, e.g. UO_6^{6-} units in β-U_3O_8 adopt a tetragonally distorted octahedral coordination with shortened axial O–U–O distances that are best viewed as $[(UO_2)O_4]^{6-}$. (See Chapter 3.)

There are also several uranium imido complexes that are isoelectronic with $[UO_2]^{2+}$ and these will be considered in Section 4.10.2.

The linear $[UO_2]^{2+}$ ion shows an intense IR absorption at ca. 930 cm^{-1} due to the asymmetric O–U–O stretch.

4.2.5 Kinetics of water exchange for lanthanoid and actinoid aqua ions

Lanthanoid aqua ions are all extremely labile, but there is a pronounced variation in first order rate constants for water exchange along the series as ionic radius decreases. Figure 4.7 shows an increase in rate constant from La^{3+} to Sm^{3+} followed by a decrease from Gd^{3+} to Lu^{3+}. The origin of this variation lies in the coordination numbers of the aqua ions: nine-coordinate for the early Ln and eight-coordinate for the late Ln.

$[Ln(H_2O)_9]^{3+}$ undergo water exchange via a dissociative I_d mechanism with an eight-coordinate transition state: this is because formation of a ten-coordinate transition state in an associative mechanism would be sterically highly unfavourable. $[Ln(H_2O)_8]^{3+}$ adopt an associative I_a mechanism for water exchange, with a nine-coordinate transition state. The rates of water exchange for a particular Ln therefore depend on the relative stabilities of $[Ln(H_2O)_9]^{3+}$ and $[Ln(H_2O)_8]^{3+}$ for that Ln. For La^{3+} to Nd^{3+}, the nine-coordinate aqua ion is more stable than the eight-coordinate ion, with the difference in relative stability decreasing as Ln^{3+} radius decreases. The rate of water exchange thus increases from La^{3+} to Nd^{3+}.

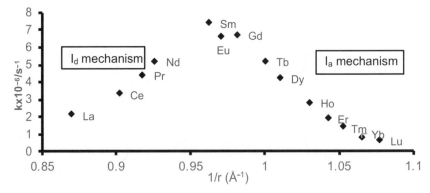

Figure 4.7 Water exchange kinetics for $[Ln(H_2O)_n]^{3+}$

The situation is reversed for the later and smaller Gd^{3+} to Lu^{3+}, in which the eight-coordinate aqua ion becomes progressively more stable with decreasing Ln^{3+} radius, and the rate of water exchange decreases correspondingly. For Sm^{3+} and Eu^{3+}, there is very little difference in stability between eight- and nine-coordinate aqua ions, and the rate constant for water exchange is at a maximum for these two lanthanoids.

Eu^{2+} is accessible in aqueous solution and it has been shown by EXAFS to exist as an equilibrium mixture of predominantly $[Eu(H_2O)_7]^{2+}$ and $[Eu(H_2O)_8]^{2+}$ with an average of 7.2 water ligands per Eu. The water exchange rate of $[Eu(H_2O)_7]^{2+}$ is faster than for any of the Ln^{3+} aqua ions, with a rate constant of 5×10^9 s^{-1}, and the mechanism of this exchange reaction is thought to be associative.

The +4 oxidation state is important in aqueous solution for Th, Pa, U, and Pu. Due to the high charge density of the An^{4+} ions, their aqua complexes are strong Brønsted acids which are susceptible to hydrolysis at pH > 2, forming oxo- and hydroxo-bridged oligomers and polymers, and ultimately highly insoluble hydrated oxides. However, under acidic conditions they exist as $[An(H_2O)_n]^{4+}$ with n = 10±1. Water exchange for $[U(H_2O)_{10}]^{4+}$ is believed to occur via a *D* mechanism and has a rate constant of approximately 5×10^6 s^{-1}.

Water exchange for $[UO_2(H_2O)_5]^{2+}$ is slightly slower than for $[U(H_2O)_{10}]^{4+}$ with a first order rate constant of 1.3×10^6 s^{-1}. The mechanism is proposed to be *I* or *A*.

4.3 Kinetics of complex formation in aqueous solution

As with water exchange, the kinetics of complex formation in aqueous solution is influenced by Ln^{3+} radius. Figure 4.8 shows a plot of rate constant vs. 1/r for oxalate complexation with Ln^{3+}. The rate constant remains almost unchanged from La^{3+} to Eu^{3+} then decreases from Gd^{3+} to Dy^{3+} and remains almost constant from Dy^{3+} to Tm^{3+}. These observations are explained by the change in coordination number of aqua ions from nine for the early Ln^{3+} to eight for the smaller late Ln^{3+}, with a corresponding decrease in Ln–O distance and increase in Ln–O bond

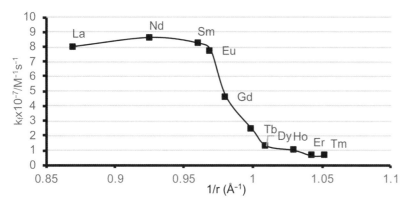

Figure 4.8 Rate constants for formation of 1:1 complexes of Ln^{3+} with oxalate $C_2O_4^{2-}$ plotted against 6-coordinate Ln^{3+} radius

strength for inner sphere H_2O from $[La(H_2O)_9]^{3+}$ to $[Tm(H_2O)_8]^{3+}$. The rate-determining step in complex formation is dissociation of inner sphere H_2O, a process that is more facile for $[La(H_2O)_9]^{3+}$ than for $[Tm(H_2O)_8]^{3+}$.

Detailed kinetic data for complexation reactions of actinoids are not available.

An inner sphere complex has a direct bond between metal ion and ligand L; in an outer sphere complex, L is loosely associated with a coordinatively saturated metal complex and there is no direct metal to L bond.

4.4 Thermodynamics of complex formation in aqueous solution

This discussion will focus mainly on studies of Ln coordination chemistry, for which data are more complete than for An.

Hydration enthalpies for Ln^{3+} are very large (see Figure 4.4), and so the process of substituting coordinated H_2O with an incoming ligand is often endothermic. This means that formation of thermodynamically stable complexes (negative $\Delta_{rxn}G°$ for complex formation) in aqueous solution requires large and positive $\Delta_{rxn}S°$ and so multidentate chelating ligands are required.

Figure 4.9 shows a plot of stability constant data ($\log_{10}\beta$) for Ln^{3+}. The main points to notice are:

- Cl^- forms only outer sphere complexes with Ln^{3+} resulting in very low values (< 0.5) for $\log_{10}\beta$.
- F^- is a harder ligand than Cl^- and forms inner sphere complexes with Ln^{3+} resulting in higher values for $\log_{10}\beta$.
- $\log_{10}\beta$ increases from F^- to $DTPA^{5-}$ due to the chelate effect as the number of ligand donor atoms increases from one to eight.
- For ligands other than $DTPA^{5-}$ there is an increase in $\log_{10}\beta$ as ionic radius decreases from La^{3+} to Lu^{3+}, resulting in a higher charge:radius ratio and shorter, stronger Ln–ligand bonds.
- For $DTPA^{5-}$ there is an increase in $\log_{10}\beta$ with decreasing Ln^{3+} radius from La^{3+} to Dy^{3+}; $\log_{10}\beta$ then decreases slightly from Dy^{3+} to Lu^{3+} as strain is introduced into the ligand as it distorts to accommodate the smaller Ln^{3+} ions.

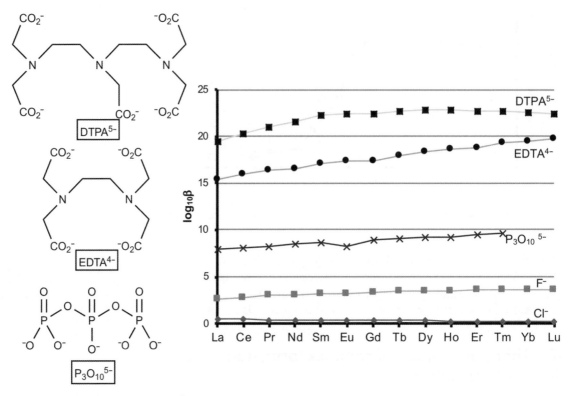

Figure 4.9 Stability constant data for Ln^{3+} complexes in aqueous solution

Stability constants for complexation of Ln^{3+} with inorganic oxo anions increase with increasing Brønsted basicity of the anion (Ln^{3+} and H$^+$ are in competition for binding) in the order:

$$NO_3^- < SO_4^{2-} \ll CO_3^{2-} < PO_4^{3-}$$

Stability constants for complexation of the hard Ln^{3+} ions with halide anions increase in order of increasing hardness of halide:

$$I^- < Br^- < Cl^- \ll F^-$$

F$^-$ is the only halide that forms an inner sphere complex with Ln^{3+}: the other halide ions only ever become loosely associated with the second coordination sphere.

The overwhelming importance of the entropy contribution to complexation in water is shown in the plot of $\Delta H°$, $T\Delta S°$, and $\Delta G°$ for complexation of Ln^{3+} with EDTA^{4-} (Figure 4.10). It is also important to note that neither $\Delta H°$ nor $T\Delta S°$ varies monotonically along the lanthanoid series: irregularities are explained by subtle effects of solvation including the change in coordination number of aqua ions around the middle of the series.

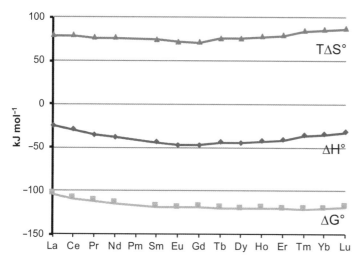

Figure 4.10 Thermodynamic parameters for Ln^{3+} complexation with $EDTA^{4-}$ in aqueous solution at 25 °C

4.5 Complexes that are thermodynamically stable in aqueous solution

It is clear from Figure 4.9 that chelating ligands are essential in order to form complexes that are thermodynamically stable in aqueous solution.

Metal complexes with bidentate (or multidentate) chelating ligands are more thermodynamically stable than analogous complexes with an equivalent number of monodentate ligands. This is known as the chelate effect and is due to generally favourable entropy contributions on formation of chelate complexes. For example, when one bidentate ligand LL displaces two H_2O ligands as shown in equation (1), there is a net increase in independent particles (two on the left vs. three on the right). However, when two equivalent monodentate ligand molecules L displace two H_2O ligands (equation (2)) there is no net increase in independent particles (three on the left vs. three on the right).

$$[Ln(H_2O)_n]^{3+} + LL \rightarrow [Ln(H_2O)_{n-2}(LL)]^{3+} + 2H_2O \quad (1)$$

$$[Ln(H_2O)_n]^{3+} + 2L \rightarrow [Ln(H_2O)_{n-2}L_2]^{3+} + 2H_2O \quad (2)$$

Polyaminocarboxylate ligands such as $EDTA^{4-}$ and $DTPA^{5-}$ form stable complexes with all the f-elements, and complexes with An^{3+} are analogous to those of Ln^{3+}. Neither $EDTA^{4-}$ nor $DTPA^{5-}$ has sufficient donor atoms to fill the coordination spheres of the large Ln/An^{3+} ions, and so the complexes retain coordinated H_2O ligands, e.g. $[Ln(EDTA)(H_2O)_3]^-$ and $[Ln(DTPA)(H_2O)]^{2-}$.

The macrocyclic $DOTA^{4-}$ ligand (Figure 4.11) forms complexes with Ln^{3+} that are even more stable than those with $DTPA^{5-}$: e.g. $\log_{10}\beta \approx 28$ for $[Gd(DOTA)(H_2O)]^-$. Complexes of Gd^{3+} with polyaminocarboxylate ligands are used in diagnostic medicine as contrast agents for MRI imaging (see Chapter 6).

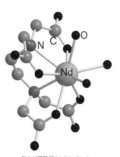

$[Nd(EDTA)(H_2O)_3]^-$

Table 4.1 Stability constants for 1:1 complexes of Pu with $EDTA^{4-}$

Oxidation state	+5	+4	+3
Pu ion	PuO_2^+	Pu^{4+}	Pu^{3+}
$\log_{10}\beta$	12.3	26.4	16.1

The early An may have several accessible oxidation states giving rise to a range of complexes. As an example, Table 4.1 summarizes stability constants for Pu complexes with $EDTA^{4-}$.

The stability constants follow the expected order based on the effective charge Z_{eff} at Pu.

Actinoid ions are highly toxic due to the radiological damage they can cause, and Pu^{4+} is additionally toxic as it can replace Fe^{3+} in transferrin and ferritin, so there has been considerable effort aimed at designing sequestering agents to treat An poisoning. The multidentate O,N-donor ligand 3,4,3-LI(1,2-HOPO) (Figure 4.11) has higher stability constants for An binding than $DTPA^{5-}$ and is effective for Pu^{4+} sequestration.

Actinyl ions $[AnO_2(H_2O)_5]^{x+}$ have two inert axial An=O bonds and five labile $Ln-OH_2$ bonds in the equatorial plane. The equatorial H_2O ligands can be substituted in aqueous solution to form complexes in which the $AnO_2^{+/2+}$

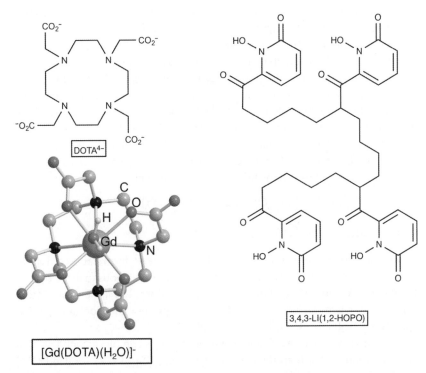

Figure 4.11 Ligands that form thermodynamically stable complexes with f-element ions in aqueous solution

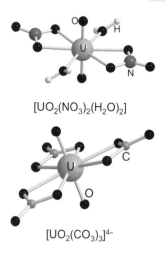

$[UO_2(NO_3)_2(H_2O)_2]$

$[UO_2(CO_3)_3]^{4-}$

Figure 4.12 Substitution reactions of $[UO_2(H_2O)_5]^{2+}$

entity remains intact, as shown in Figure 4.12. In the case of $[AnO_2(CO_3)_3]^{4-}$ the complex is very stable ($\log_{10}\beta = 21.6$ for U; 18.5 for Pu).

4.6 Complexes that are kinetically inert in aqueous solution

Applications of the lanthanoids in medicine (e.g. MRI contrast agents) and physiology (e.g. luminescent labels) are becoming increasingly important. These applications require complexes that are inert under physiological conditions (i.e. aqueous solution at pH ≈ 7), and as we have seen, there are a rather limited number of ligands that form thermodynamically stable complexes with lanthanoids in aqueous solution. However, some complexes are kinetically inert in aqueous solution, but due to low stability constants in water, they must be prepared in non-aqueous solution.

4.6.1 Schiff base macrocycle complexes

Metal ion templated Schiff base condensation reactions provide a route into numerous macrocycle complexes of the lanthanoids, many of which are kinetically inert in water (Figure 4.13).

There are almost unlimited possibilities for the di-carbonyl and diamine starting materials resulting in an enormous number of possible ligand structures. Use of chiral diamines gives a route into chiral complexes that have been explored for applications in enantioselective catalysis.

4.6.2 Cryptate complexes

Macrobicyclic cryptand ligands such as (2.2.2) and bipy(2.2.2) shown in Figure 4.14 have pre-organized cavities that can accommodate Ln^{3+} ions to form cryptate complexes that are soluble and remarkably inert in water. However, the formation constants in water are not high enough to allow synthesis in aqueous solution, and these complexes must be prepared in strictly anhydrous conditions.

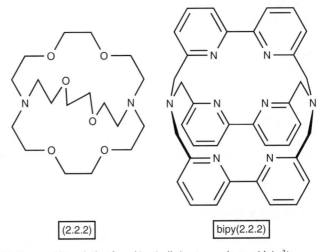

Figure 4.13 Schiff base condensation reaction to form an Ln^{3+} complex with a macrocyclic ligand

(2.2.2)   bipy(2.2.2)

Figure 4.14 Cryptand ligands that form kinetically inert complexes with Ln^{3+}

$[Eu(2.2.2)(NO_3)]^{2+}$

Complexes of Eu^{3+} and Tb^{3+} with bipy(2.2.2) display highly efficient luminescence in aqueous solution (see Chapter 2) and are commercially available as fluorescent labels for high throughput drug target studies (see Chapter 6).

4.7 Complexes with neutral O-donor ligands

4.7.1 P=O donors: phosphine oxides and phosphates

The P=O functional group is a very strong donor, and there is an extensive coordination chemistry of the f-elements with phosphine oxides $R_3P=O$ and organophosphates $(RO)_3P=O$. Much of this chemistry relates to the use of these ligands in solvent extraction techniques for extraction and purification of the metals. For example the complex of uranyl nitrate with tri-n-butylphosphate $[UO_2(NO_3)_2$ $\{(Bu^nO)_3P=O\}_2]$ is crucial in the PUREX process for reprocessing nuclear fuels (see Chapter 7). P=O donors are also used in combination with other ligands (e.g. β-diketonates; Section 4.9.1) to modify properties such as stability and luminescence efficiency. HMPA $((Me_2N)_3P=O)$ is used to increase the reducing power of SmI_2 (see Chapter 6).

4.7.2 Polyether ligands: crown ethers and podands

The coordination chemistry of the lanthanoids with crown ethers and their acyclic podand analogues has been investigated to establish the effect of Ln^{3+} radius on stability constants, in part motivated by the search for reagents for selective extraction of Ln^{3+} from mixtures. None of the complexes of these ligands with Ln^{3+} is sufficiently stable to exist in aqueous solution, so the complexes must all be prepared in non-aqueous solution, sometimes under very rigorously anhydrous conditions.

Figure 4.15 shows plots of stability constants $(\log_{10}\beta)$ for complexation of $Ln(OTf)_3$ in propylenecarbonate (donor number 15.1) with crown ethers and podands.

$[UO_2(NO_3)_2\{(MeO)_3P\}_2]$

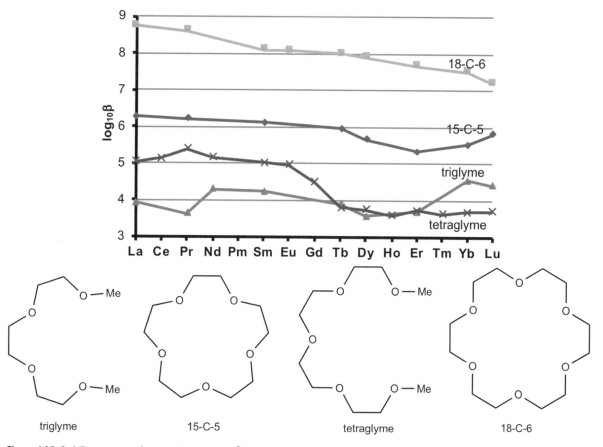

Figure 4.15 Stability constants for complexation of Ln^{3+} with crown ether and podand ligands (measured in propylenecarbonate)

For all Ln^{3+} there is a clear trend of increased stability for complexes with 15-C-5 compared with its acyclic analogue tetraglyme, demonstrating the macrocyclic effect. For the crown ether ligands there is an increase in stability with increased number of donor atoms so that for all Ln^{3+}, complexes with 18-C-6 are at least an order of magnitude more stable than their 15-C-5 analogues. Trends in stability constant along the Ln^{3+} series are more subtle: there is a steady decrease in stability of 18-C-6 complexes with decreasing Ln^{3+} radius as the ligand has to distort to bind to the smaller ions, as shown in Figure 4.16 for complexes of La^{3+} and Dy^{3+}. Tetraglyme shows a maximum stability for complexation with Pr^{3+}, and a significant decrease in stability for the later lanthanoids Tb^{3+}–Lu^{3+}. This decrease is due to unfavourable steric interactions between OMe groups as tetraglyme binds to the smaller Ln^{3+}.

Actinyl ions [NpO$_2$]$^+$ and [UO$_2$]$^{2+}$ form complexes with 18-C-6; [NpO$_2$(18-C-6)]$^+$ can form in aqueous solution whereas [UO$_2$(18-C-6)]$^{2+}$ only exists in non-aqueous solution. There is no strong evidence that this difference is due to a better fit of [NpO$_2$]$^+$ with 18-C-6 compared with [UO$_2$]$^{2+}$: it has been explained by the difference in magnitude of $\Delta_{hyd}H°$, which is smaller for [NpO$_2$]$^+$ than for [UO$_2$]$^{2+}$.

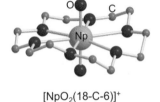

[NpO$_2$(18-C-6)]$^+$

[La(NO$_3$)$_3$(18-C-6)]   [Eu(NO$_3$)$_3$(15-C-5)]

[DyCl(18-C-6)(H$_2$O)$_2$]$^+$ [La(NO$_3$)$_3$(tetraglyme)]

Figure 4.16 Structures of Ln^{3+} complexes with crown ether and podand ligands

4.8 Complexes with neutral N-donor ligands

4.8.1 Aliphatic amines

Amines are far too basic to be useful ligands for f-elements in aqueous solution: for en, $pK_b \approx 4$ and so this would just result in precipitation of metal hydroxide from solution. However, the thermodynamics of aliphatic amine coordination to Ln^{3+} in DMSO solution have been investigated in detail. The monodentate primary amine Bu^nNH_2 does not form inner sphere complexes with Ln^{3+} in DMSO, but chelating ligands such as en, dien, trien, and tren do form complexes in this solvent.

Figure 4.17 shows $\log_{10}\beta$ values for $[LnL]^{3+}$. As expected there is an increase of over 2 orders of magnitude on increasing the number of donor atoms from 2 to 4, and there is also the expected increase with decreasing Ln^{3+} radius as the electrostatic ligand–Ln interactions become stronger.

A more thorough consideration of thermodynamic factors shows that there are subtle solvent-dependent effects at play: inspection of $\Delta H°$ and $T\Delta S°$ values (see Figure 4.18) shows that for all Ln^{3+} $\Delta S°$ is negative, and complex formation with en (and the other chelating amines shown in Figure 4.17) in DMSO is an enthalpy-driven process.

Complexation of chelating alkylamines to Ln^{3+} contrasts with the complexation of $EDTA^{4-}$ to Ln^{3+} in aqueous solution, which is entropy-driven

4.8.2 Aromatic N-donors

There is an extensive coordination chemistry of the f-elements with aromatic N-donors such as bipy, phen, and terpy (see Figure 4.19). Historically these complexes laid important foundations of f-element coordination chemistry and helped to establish the principles of coordination numbers and geometries. They are frequently used as 'antenna' ligands in the preparation of luminescent Ln complexes (see Chapter 2 and Section 4.9.1).

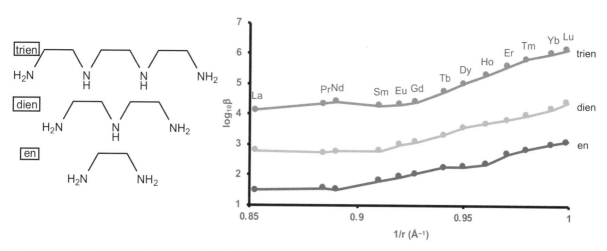

Figure 4.17 Stability constants for complexation of Ln^{3+} with amine ligands (measured in DMSO)

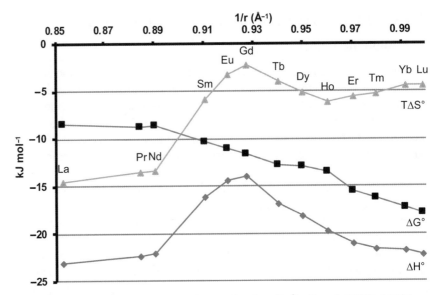

Figure 4.18 Thermodynamic parameters for complexation of Ln^{3+} with en in DMSO at 25 °C

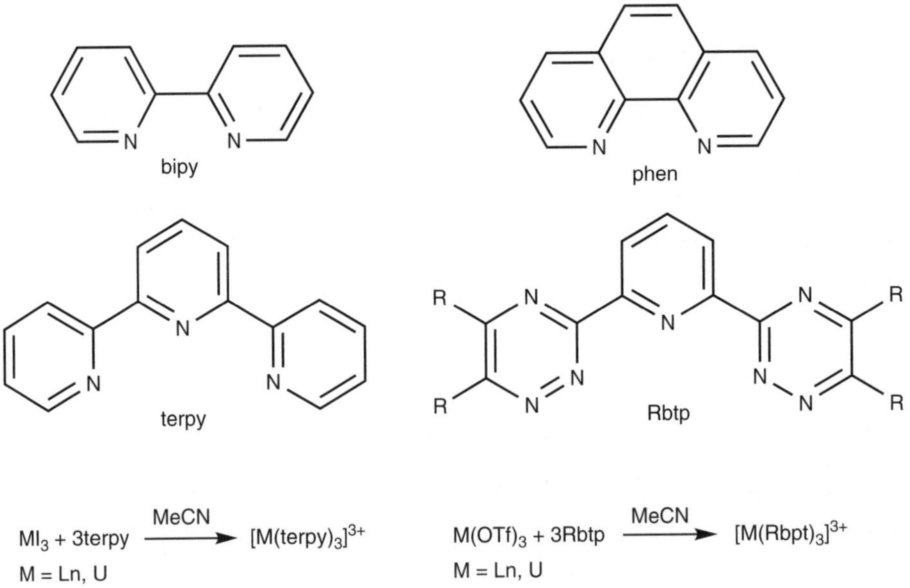

bipy

phen

terpy

Rbtp

$MI_3 + 3terpy \xrightarrow{\text{MeCN}} [M(terpy)_3]^{3+}$

$M = Ln, U$

$M(OTf)_3 + 3Rbtp \xrightarrow{\text{MeCN}} [M(Rbpt)_3]^{3+}$

$M = Ln, U$

Figure 4.19 Examples of aromatic N-donors and their complexes with Ln^{3+} and U^{3+}

9-coordinate complexes $[ML_3]^{3+}$ (L = terpy or Rbpt) are formed by reaction of MI_3 or $M(OTf)_3$ (M = Ln, U) with 3 equiv of L in MeCN as shown in Figure 4.19. In competition reactions, both terpy and Rbpt show a greater affinity for U^{3+} than for Ln^{3+}; the effect is greater for Rbpt, and could be exploited in Ln^{3+}/An^{3+} separations as part of the processing of nuclear waste. Comparison of M–N distances (Table 4.2) shows that in both terpy and Rbpt complexes, the average

Table 4.2 Structural data for $[M(terpy)_3]^{3+}$ and $[M(Rbpt)_3]^{3+}$

M	M^{3+} radius (6-coord.)/pm	Ave. M–N distance/pm $[M(terpy)_3]^{3+}$	Ave. M–N distance/pm $[M(Rbpt)_3]^{3+}$ (R = Pr)
Ce	115	265	263
U	116	262	255

U–N distance is shorter than the average Ce–N distance, despite the fact that the radius of U^{3+} is slightly larger than that of Ce^{3+}. This has been interpreted as evidence for π-backbonding between the Rbpt ligand and U^{3+}, which is not possible for Ln^{3+}.

4.9 Complexes with anionic O-donor ligands

4.9.1 β-diketonates

Lanthanoid β-diketonates have been known since the early years of the twentieth century. Systematic investigations of their chemistry began in the late 1940s, and continue in the twenty-first century mainly due to their luminescent properties and their applications as precursors in materials synthesis. They can be synthesized straightforwardly by a salt exchange reaction as shown in Figure 4.20.

The complexes display varying structures depending on the steric demands of the β-diketonate ligand and the Ln^{3+} radius as shown in Figure 4.21. For example $[La(acac)_3]$ adopts a polymeric structure in the solid state whereas complexes with the more sterically demanding thd ligand are dimeric for large Ln^{3+} (e.g. Pr^{3+}) and monomeric for small Ln (e.g. Yb^{3+}). Anionic tetrakis(β-diketonate) complexes are known for the less sterically demanding ligands such as acac, dbm, and hfa.

Reactivity of $[Ln(diket)_3]$ is really limited to adduct formation with Lewis bases; many adducts with strong donors such as pyridine or Ph_3PO have been characterized by x-ray diffraction. The number of donor ligands is determined by the steric requirements of diket and of the Lewis base; although they are not isolable, labile adducts with alcohols and ethers are observed in solution. Apart from formation of H_2O adducts $[Ln(diket)_3]$ are relatively unreactive with water and may be handled in air without decomposition.

Apart from polymeric $[Ln(acac)_3]$ all the lanthanoid β-diketonates are soluble in organic solvents, and several are volatile. In the 1960s, GC separation of

ligand name	R
acac	Me
thd	But
hfa	CF3
dbm	Ph

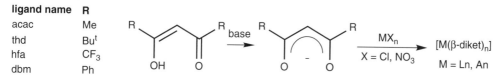

Figure 4.20 Synthesis of β-diketonate complexes

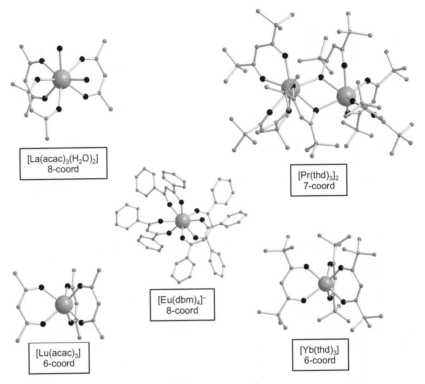

Figure 4.21 Structures of lanthanoid β-diketonate complexes

[Ln(diket)$_3$] was considered as a means of separating individual Ln. This is not of current practical interest, but volatile [Ln(diket)$_3$] complexes are now of interest as precursors for synthesis of thin films of Ln oxides by MOCVD (Metal Organic Chemical Vapour Deposition).

The photophysics of lanthanoid β-diketonates have been well-studied: sensitized emission from Eu complexes under UV irradiation was first observed in 1942, and a present-day application of this effect is the use of a Eu bis(β-diketonate) complex in the security ink on Euro banknotes. The European Central Bank has not disclosed its exact identity.

Adducts of [Eu(diket)$_3$] with P=O donors or aromatic N-donors (see Figure 4.22) show enhanced stability and luminescent properties compared with the parent [Eu(diket)$_3$].

Some lanthanoid β-diketonate complexes show the intriguing property of triboluminescence. When crystals of [HNEt$_3$][Eu(dbm)$_4$] are crushed they emit bright orange-red flashes characteristic of Eu^{3+} emission, but only if they have been crystallized from the correct solvent: disordered crystals from MeOH are triboluminescent whereas crystals of the CH$_2$Cl$_2$ solvate are not. [HNEt$_3$][Eu(dbm)$_4$] and related complexes have been investigated as sensors for monitoring mechanical stress, e.g. in aircraft wings.

[Eu(tta)$_3$(dpepo)]   [Eu(dbm)$_3$(phen)]

Figure 4.22 Luminescent adducts of [Eu(diket)$_3$]

4.9.2 Alkoxides and aryloxides

Lanthanoid alkoxides have been known since 1958, when they were mainly of curiosity value, but their potential as precursors for synthesis of highly pure lanthanoid oxides has led to a resurgence of interest in these compounds since the early 1980s.

The large size of the Ln^{3+} ions results in lanthanoid alkoxides adopting a variety of structures dictated by the demands of the alkoxide ligand. [Ln(OR)$_3$] where R = Me, Et or Ph are involatile and virtually insoluble in benzene; they have polymeric structures where the lanthanoid coordination number is increased by formation of Ln–OR–Ln bridges. Increasing the steric bulk of the alkoxide ligand leads to decreasing degrees of oligomerization, as shown in Figure 4.23. With the very bulky O-(2,6-Bu$^t_2$-C$_6$H$_3$) ligand a three-coordinate monomeric complex is formed. Its trigonal pyramidal structure in the solid state is very similar to that adopted by other three-coordinate f-block complexes such as the tris(silylamido) and the tris(bis(trimethylsilyl)methyl) complexes. Adduct formation with small Lewis bases can occur and the five coordinate [Ce(O-2,6-Me$_2$C$_6$H$_3$)$_3$(RCN)$_2$] has been reported.

Lanthanoid alkoxide chemistry is inevitably dominated by the +3 oxidation state. However, as expected, the hard alkoxide ligand can stabilize high oxidation states, and complexes of Ce(IV) with OBu^t and OPr^i are well known. These complexes are synthesized by the salt exchange reactions of [NH$_4$]$_2$[Ce(NO$_3$)$_6$] with NaOR as shown in Figure 4.24.

The +2 oxidation state is very accessible for Sm, Eu, and Yb, and there are many examples of Ln(II) alkoxides/aryloxides for these elements (e.g. see Figure 4.24). The large radii of Ln^{2+} ions compared with Ln^{3+} (e.g. 6-coord. radii: Eu^{2+} = 1.31 Å; Eu^{3+} = 1.09 Å) require a combination of bulky ligands and increased coordination numbers to satisfy their coordination requirements.

As expected, alkoxides of the early actinoids are known in a range of oxidation states as illustrated in Figure 4.25 for ethoxides of uranium.

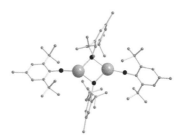

[Yb(O-2,6-Me$_2$-4-Me-C$_6$H$_2$)$_2$]$_2$

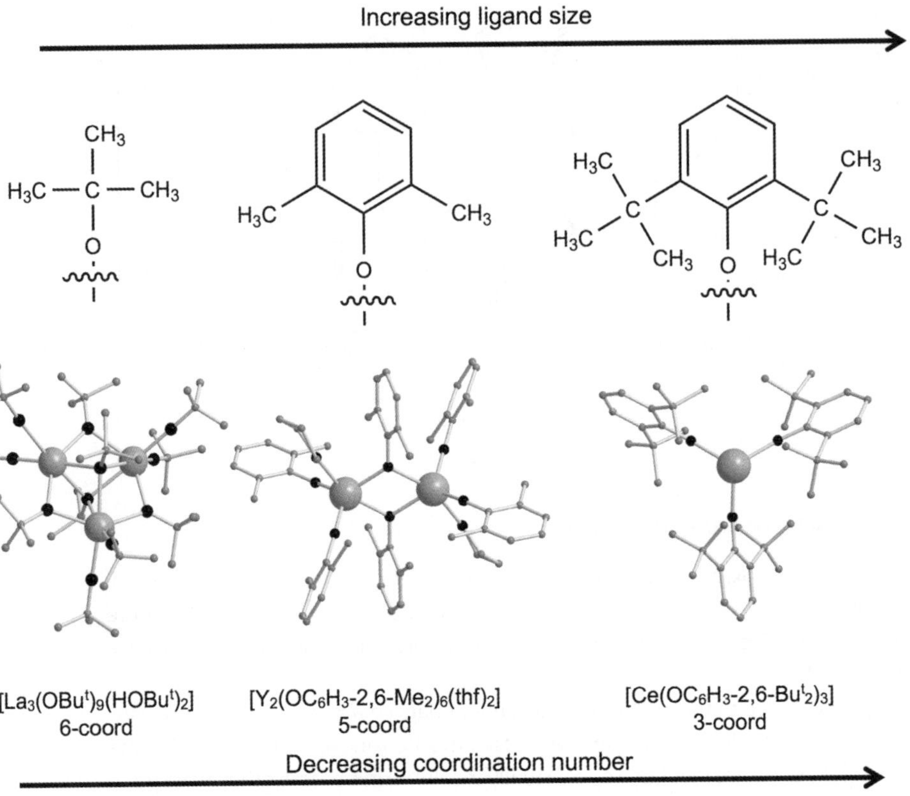

Increasing ligand size

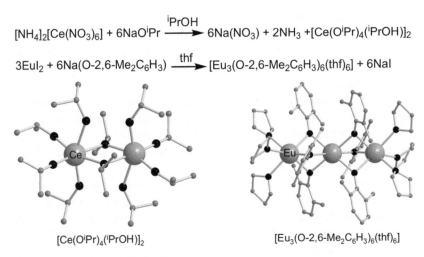

$[La_3(OBu^t)_9(HOBu^t)_2]$
6-coord

$[Y_2(OC_6H_3-2,6-Me_2)_6(thf)_2]$
5-coord

$[Ce(OC_6H_3-2,6-Bu^t_2)_3]$
3-coord

Decreasing coordination number

Figure 4.23 Lanthanoid(III) alkoxides: effect of ligand size on coordination number

$$[NH_4]_2[Ce(NO_3)_6] + 6NaO^iPr \xrightarrow{\ ^iPrOH\ } 6Na(NO_3) + 2NH_3 + [Ce(O^iPr)_4(^iPrOH)]_2$$

$$3EuI_2 + 6Na(O-2,6-Me_2C_6H_3) \xrightarrow{\ thf\ } [Eu_3(O-2,6-Me_2C_6H_3)_6(thf)_6] + 6NaI$$

$[Ce(O^iPr)_4(^iPrOH)]_2$

$[Eu_3(O-2,6-Me_2C_6H_3)_6(thf)_6]$

Figure 4.24 Ce(IV) and Eu(II) alkoxides

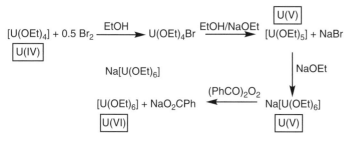

Figure 4.25 Oxidation states of uranium ethoxides

Synthesis of alkoxides

Two important synthetic routes to Ln/An alkoxides are shown in Figure 4.26. The salt exchange route is driven by the formation of stable and insoluble alkali metal halide, and the protonolysis route is driven by the more powerful Brønsted acidity of ROH compared with HNR_2.

Reactivity of alkoxides

The reactivity of f-element alkoxides is dominated by their Brønsted basicity: they react readily with protic reagents that are more powerful Brønsted acids than the parent alcohol, e.g. H_2O, β-diketones, silanols, and carboxylic acids (Figure 4.27).

Many f-element alkoxides are Lewis acidic and adduct formation with Lewis bases is another common reaction. When the synthesis is performed in a coordinating solvent such as thf, a solvated complex is often formed in order to achieve coordinative saturation at the metal centre.

Ln alkoxides in catalysis

The combination of Brønsted basicity and Lewis acidity has been exploited in catalytic reactions. A particularly elegant example of this bifunctional catalysis is the use of $[M]_3[Ln(binol)_3]$ (binol = 2,2-binaphtholate) to catalyse enantioselective nitroaldol reactions as shown in Figure 4.28. This reaction is highly sensitive to Ln^{3+} radius and to the identity of M.

$$Ln/AnX_n + nMOR \xrightarrow[\text{anhydrous}]{\text{thf or } Et_2O} [Ln/An(OR)_n] + nMX \quad \text{salt exchange}$$

M = Li, Na, K
X = halide

$$[Ln/An(NR_2)_n] + nROH \longrightarrow [Ln/An(OR)_n] + nR_2NH \quad \text{protonolysis}$$

Figure 4.26 Synthesis of f-element alkoxides/aryloxides

$$[Ln/An(OR)_n] + nHX \longrightarrow [Ln/AnX_n] + nHOR$$

Figure 4.27 Protonolysis reaction of f-element alkoxides

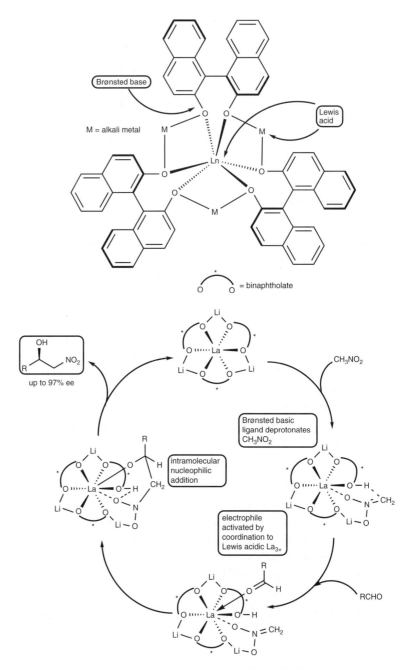

Figure 4.28 Enantioselective nitroaldol reaction catalysed by $M_3[Ln(binol)_3]$

Polylactide (also known as polylactic acid or PLA) is by volume the second most important bioplastic in the world. The Lewis acidity of Ln alkoxides, combined with the lability of the Ln–OR bond results in many Ln alkoxides being highly active catalysts for the living ring-opening polymerization of lactide to form biodegradable polylactide as shown in Figure 4.29.

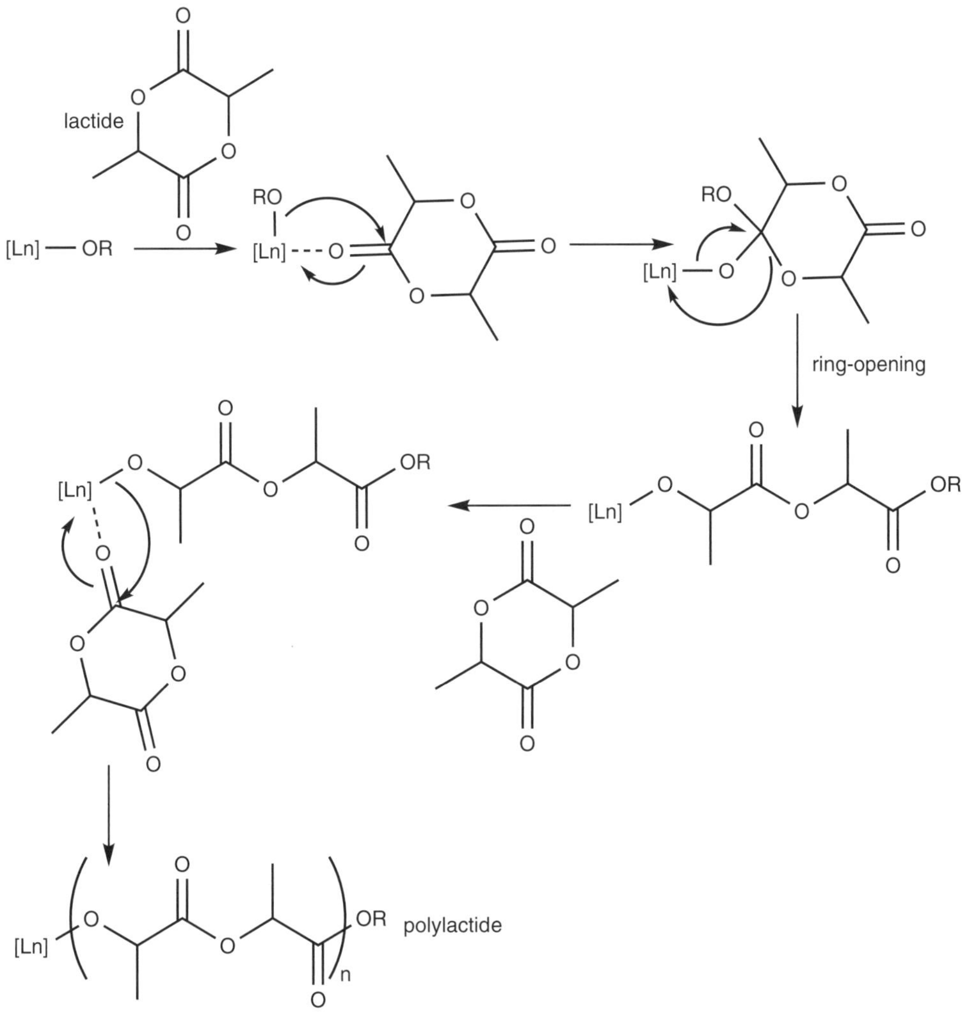

Figure 4.29 Living ring-opening polymerization of lactide catalysed by Ln alkoxide ([Ln] = Ln + ancillary ligands)

4.10 Complexes with anionic N-donors

4.10.1 Silylamides and alkylamides

The silylamide ligand ($N(SiMe_3)_2^-$) occupies a special place in f-element coordination chemistry: it is very sterically demanding and enabled the synthesis of the first ever three-coordinate f-element complexes, $[Ln\{N(SiMe_3)_2\}_3]$, in 1972. The synthesis was achieved by a salt exchange reaction of anhydrous $LnCl_3$ with $LiN(SiMe_3)_2$ under strictly anhydrous conditions. The U(III) analogue was synthesized in 1979, starting from UCl_4 in the presence of a reducing agent. Figure 4.30 summarizes syntheses of a range of lanthanoid and actinoid silylamides and shows how the silylamide ligand is able to stabilize oxidation states from +2 to +6.

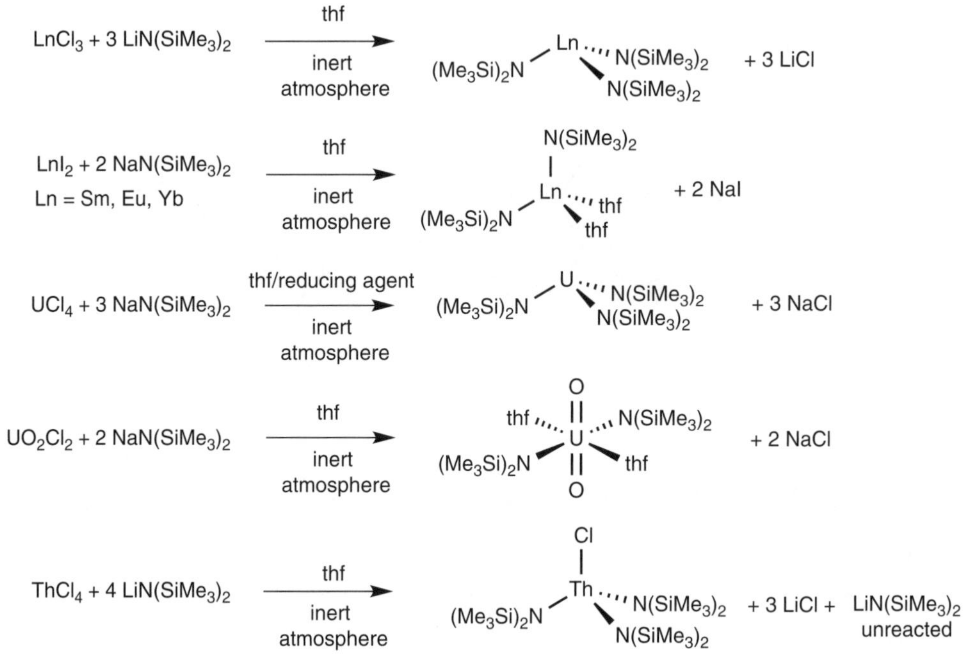

LnCl3 + 3 LiN(SiMe3)2
thf
inert atmosphere
(Me3Si)2N
Ln
N(SiMe3)2
N(SiMe3)2
+ 3 LiCl
LnI2 + 2 NaN(SiMe3)2
Ln = Sm, Eu, Yb
thf
inert atmosphere
N(SiMe3)2
(Me3Si)2N
Ln
thf
thf
+ 2 NaI
UCl4 + 3 NaN(SiMe3)2
thf/reducing agent
inert atmosphere
(Me3Si)2N
U
N(SiMe3)2
N(SiMe3)2
+ 3 NaCl
UO2Cl2 + 2 NaN(SiMe3)2
thf
inert atmosphere
O
thf
U
N(SiMe3)2
(Me3Si)2N
thf
O
+ 2 NaCl
ThCl4 + 4 LiN(SiMe3)2
thf
inert atmosphere
Cl
Th
(Me3Si)2N
N(SiMe3)2
N(SiMe3)2
+ 3 LiCl +
LiN(SiMe3)2
unreacted

Figure 4.30 Synthesis of f-element silylamides

(Me3Si)2N
Ln
N(SiMe3)2
N(SiMe3)2
Ph3P=O
PPh3
O
Ln
(Me3Si)2N
N(SiMe3)2
N(SiMe3)2
Ph3P=O
PPh3
O
(Me3Si)2N
(Me3Si)2N
Ln
O
O
Ln
N(SiMe3)2
N(SiMe3)2
O
PPh3

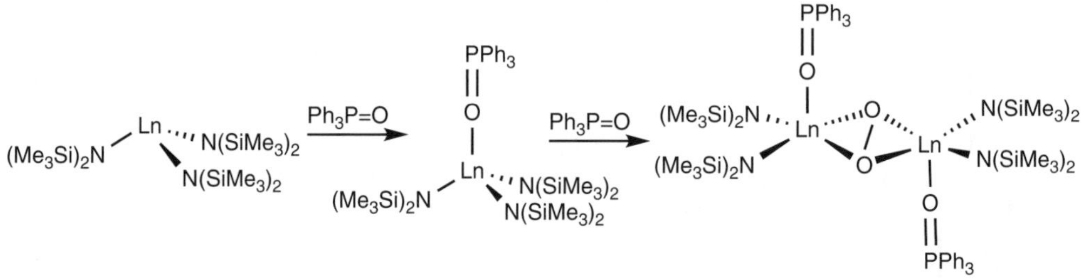

Figure 4.31 Reactions of $[Ln\{N(SiMe_3)_2\}_3]$ with $Ph_3P{=}O$

Si
N

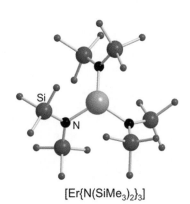

$[Er\{N(SiMe_3)_2\}_3]$

The structures of $[M\{N(SiMe_3)_2\}_3]$ (M = Ln or An) are all pyramidal in both the solid state and the gas phase: the pyramidal structure is stabilized by forming a dipole between the N_3^{3-} plane and the M^{3+} ion, and in the solid state there is additional stabilization through intramolecular van der Waals forces.

Two factors dominate the reactivity of f-element silylamides: Lewis acidity at the metal centre and Brønsted basicity of the ligand. Lewis acidity at the metal is demonstrated by adduct formation with Lewis bases such as thf, MeCN, and phosphine oxides, $R_3P{=}O$ (see Exercise 3 in Chapter 2). In the case of $Ph_3P{=}O$, addition of excess Lewis base to $[Ln\{N(SiMe_3)_2\}_3]$ results in O-abstraction to form a unique μ_2-peroxo complex as shown in Figure 4.31.

Figure 4.32 illustrates some contrasts in reactivity between Ln and An. Addition of Me_3NO to $[Ln\{N(SiMe_3)_2\}_3]$ results in simple adduct formation with no change in oxidation state at Ln, whereas for M = U, oxidation from U(III) to

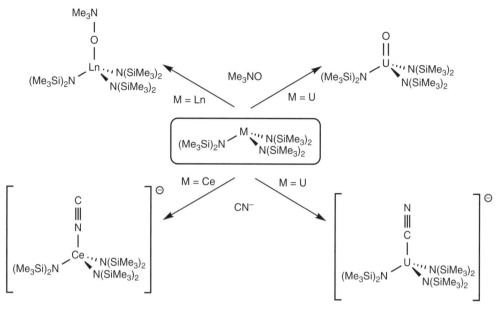

Figure 4.32 Contrasting reactivity of [Ln{N(SiMe₃)₂}₃] and [U{N(SiMe₃)₂}₃]

U(V) occurs with formation of a U=O bond. Addition of the ambidentate ligand CN⁻ to [Ce{N(SiMe₃)₂}₃] results in the hard N-donor binding to Ce; in contrast the soft C-donor binds in the U analogue.

Due to the ionic nature of the M–N bonds protonolysis reactions are important for silylamide complexes of the f-elements. The silylamide ligand is a powerful Brønsted base (pK_a for HN(SiMe₃)₂ ≈ 26 in thf) and so will react with a wide range of protic reagents including alcohols, thiols, and β-diketones. As shown in Figure 4.33, the protonolysis reaction has been used to synthesize complexes with soft S- and Se-donors as well as alkoxides and has the advantage of forming volatile amine HN(SiMe₃)₂, which can easily be removed, as the only by-product. See Section 4.11 for more examples of complexes with S-donor ligands.

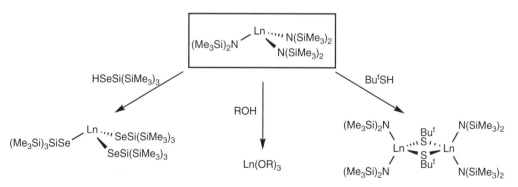

Figure 4.33 Protonolysis reactions of [Ln{N(SiMe₃)₂}₃]

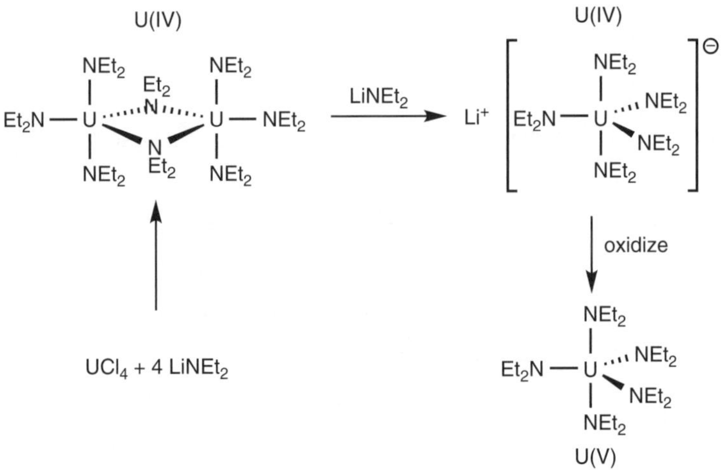

Figure 4.34 Diethylamido complexes of U(IV) and U(V)

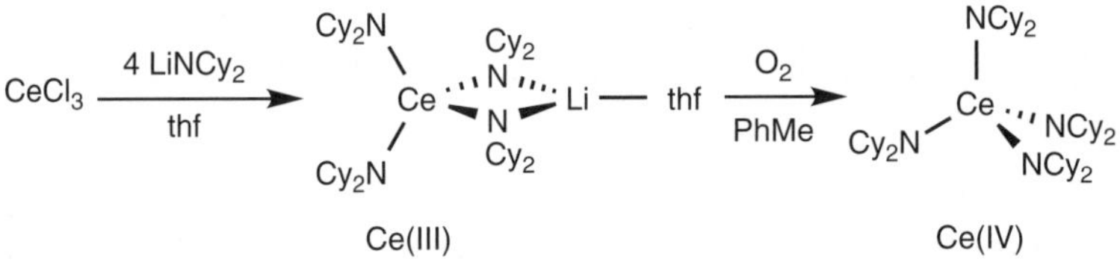

Figure 4.35 A dialkylamido complex of Ce(IV)

Dialkylamides R_2N^- are often more Brønsted basic and less sterically demanding than silylamides and they have a distinct coordination chemistry with the f-elements. For example Et_2N^- is small enough to form five coordinate complexes with U(IV) and U(V) as shown in Figure 4.34.

The first ever homoleptic Ce(IV) dialkylamide complex was prepared using dicyclohexylamide as shown in Figure 4.35. $[Ce(NCy_2)_4]$ is much more stable than the heteroleptic Ce(IV) silylamide $[Ce\{N(SiMe_3)_2\}_3Cl]$.

The increased Brønsted basicity of dialkylamides compared with silylamides means that they are more reactive as starting materials in protonation reactions.

4.10.2 Imido complexes

The imido RN^{2-} ligand is well-established in d-transition metal chemistry, but its high charge and requirement for N=M double-bonding have proved more of a challenge in f-element chemistry. However, Ln imido complexes have been isolated by the reaction scheme shown in Figure 4.36. The imido complex is stabilized by the presence of an extremely bulky tris(pyrazolyl)borate ancillary ligand, and formation of the complex is driven by elimination of CH_4. The short Lu=N distance and almost linear Lu=N–C angle are consistent with

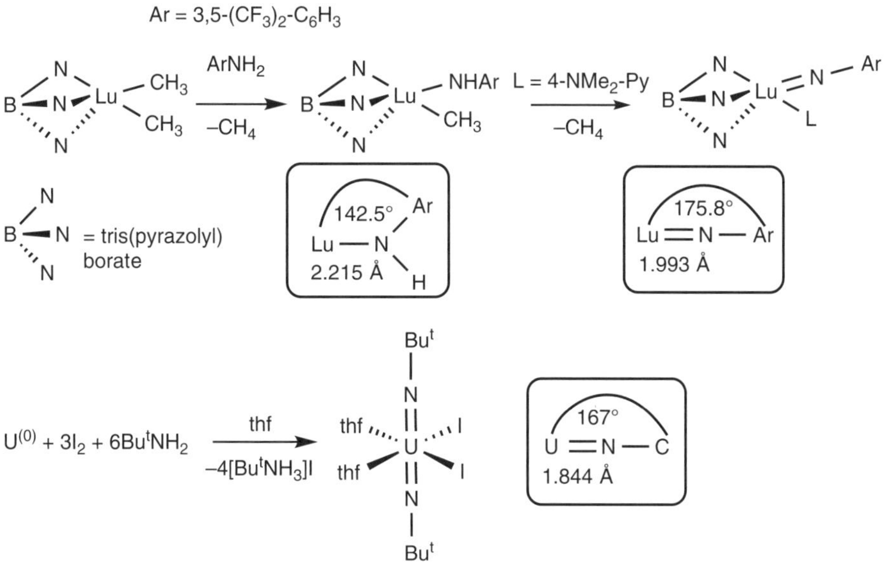

Ar = 3,5-(CF$_3$)$_2$-C$_6$H$_3$

B—N = tris(pyrazolyl) borate

Figure 4.36 Imido complexes of Lu and U

Lu=N double-bonding and an sp-hybridized N atom. Imido chemistry of U is more developed than that of Ln, and U=N bonds can be stabilized even in the absence of bulky ancillary ligands. Figure 4.36 shows an example of a U(VI) bis(imido) complex. The $[U(NR)_2]^{2+}$ motif is isoelectronic with the $[UO_2]^{2+}$ ion, and computational studies indicate that there is significant 5f and 6d contribution to U=N bonding.

4.10.3 Phthalocyanines

Lanthanoid and actinoid ions are too large to fit inside the cavity of a phthalocyanine ligand, and so they sit out of the plane of the four N-donors, forming double decker sandwich complexes with formula $[Ln(Pc)_2]^-$ or triple decker sandwiches $[Ln_2(Pc)_3]$ in which the Pc rings are parallel. $[Ln(Pc)_2]^-$ can be prepared easily by a condensation reaction of phthalonitrile with lanthanoid acetate in the presence of base as shown in Figure 4.37. $[Ln(Pc)_2]^-$ have attracted interest for applications in electrochromic devices due to their optical and redox properties, and complexes with Ln = Tb or Dy behave as single molecule magnets. (See Chapter 2.)

4.11 Complexes with soft donor ligands

Lanthanoid and actinoid ions are generally considered as hard Lewis acids and so the majority of their coordination chemistry is with hard O- and N-donor ligands. However, there is a growing coordination chemistry with softer donor

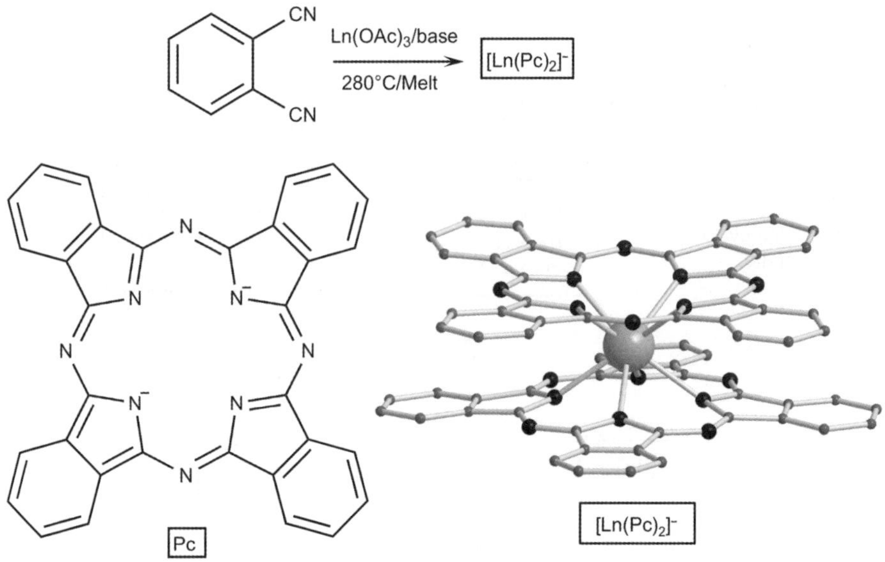

Figure 4.37 Synthesis and structure of [Ln(Pc)$_2$]$^-$ (Pc = phthalocyanine)

ligands (e.g. S- or P-donors). The possibility of covalent contributions to bonding with soft donors (especially for the early An) is of particular interest. Enhanced covalency in An complexes compared with their Ln analogues should lead to a greater affinity of S-donors for An compared with Ln, and this could be exploited in An/Ln separations that are required for reprocessing of nuclear fuels (see Chapter 7).

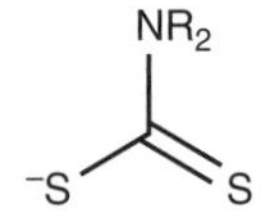

4.11.1 Group 16 donors

Dithiocarbamate ligands are planar and relatively sterically undemanding, and their solubility and electronic properties can be tuned by varying the substituents on N.

Uranyl complexes with dithiocarbamate ligands have been known since the early years of the twentieth century, and were considered for use in colorimetric determination of [UO$_2$]$^{2+}$. Since then, dithiocarbamate complexes of other An and all Ln have been prepared (Figure 4.38).

The six-coordinate complexes [Ln(S$_2$CNR$_2$)$_3$] have been investigated for applications in homogeneous catalysis and as single-source precursors for lanthanoid sulfide materials, and their adducts with aromatic N-donors such as phenanthroline display sensitized luminescence.

Dithiophosphinate ligands such as (RO)$_2$PS$_2^-$ and R$_2$PS$_2^-$ are effective for Ln(III)/An(III) separations, showing a much greater affinity for An than Ln due to enhanced covalency in An–S compared with Ln–S bonds.

The dithiophosphoramide, N(PR$_2$S)$_2^-$, ligand (see Figure 4.39) is widely used in metal extraction, and analogues with Se and Te donors have been synthesized. A detailed comparison of bonding of N(PR$_2$E)$_2^-$ (E = S, Se, or Te), with An and Ln

dithiocarbamate

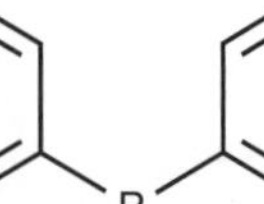

dithiophosphinate

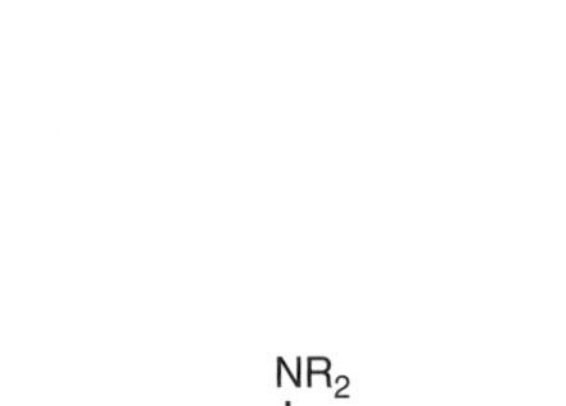

Am/Eu separation factor = 100 000

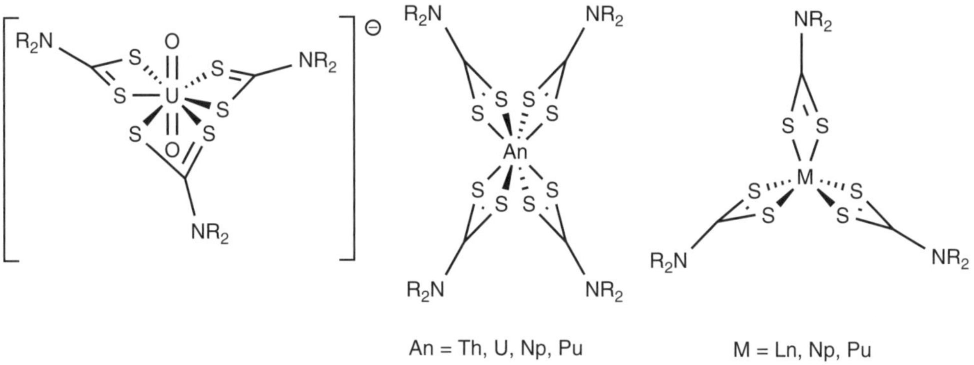

Figure 4.38 Dithiocarbamate complexes of Ln and An

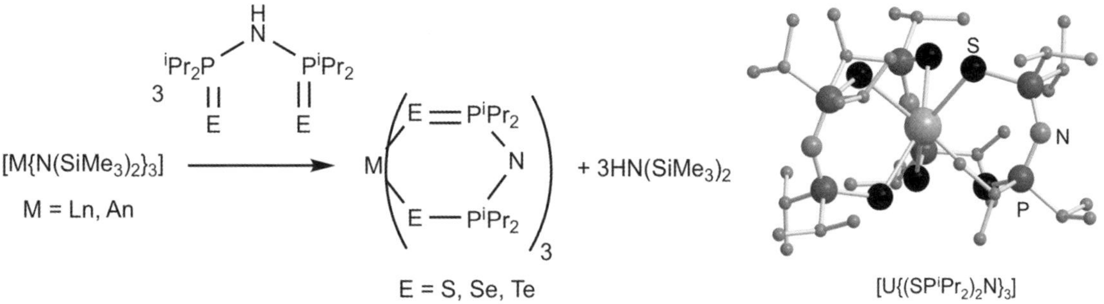

Figure 4.39 Dithiophosphoramide complexes of Ln and An

Table 4.3 Structural data for $[M\{N(P^iPr_2E)_2\}_3]$ (M = La, U; E = S, Se, Te)

	S	Se	Te
Average La–E distance/pm	289.2	301.9	322.4
Average U–E distance/pm	285.4	296.4	316.4
[La–E] – [U–E]/Å *	3.8	5.5	6.0
Δ(M–E)	3.1	4.8	5.3

* 6-coordinate radii: La^{3+} 117.2 pm; U^{3+} 116.5 pm

has been carried out. M–S distances have been measured in its six-coordinate Ln/An(III) complexes (see Table 4.3). The radius of La^{3+} is 0.7 pm greater than that of U^{3+}, but the U–S distance is 2.6 pm shorter than the La–S distance. The shorter U–S bond is consistent with a stronger bond with significant covalent contributions. The Ce^{3+}/Pu^{3+} pair shows similar trends.

4.11.2 Phosphorus donors

Tertiary phosphine, R_3P, ligands are ubiquitous in d-transition metal chemistry, where their complexes are stabilized by metal-phosphorus π-backbonding.

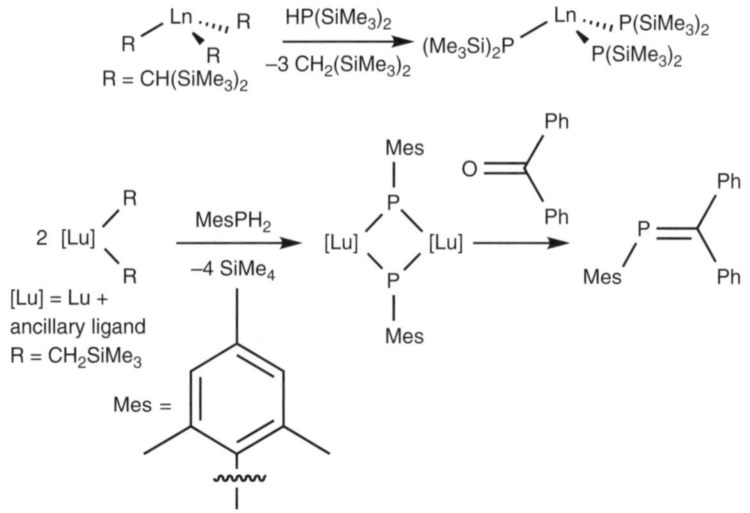

Figure 4.40 Phosphide (R_2P^-) and phosphinidene (RP^{2-}) complexes of Ln

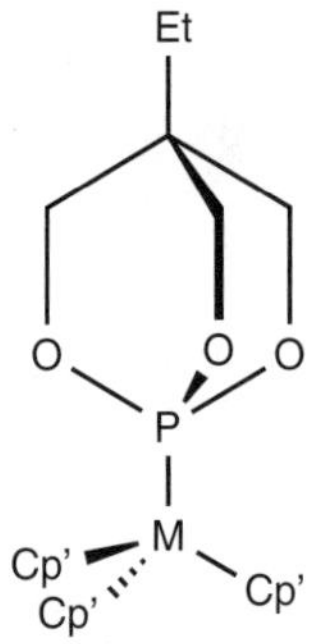

[Cp'$_3$M{P(OCH$_2$)$_3$CEt}]
M = U or Ce

Simple monodentate tertiary phosphines and phosphites will form adducts with Lewis acidic f-element centres such as [Cp$_3$M] (M = Ln or An). The P–M distances have been measured for the complexes [Cp'$_3$M{P(OCH$_2$)$_3$CEt}]. The P–U distance (2.988 Å) was found to be significantly shorter than the P–Ce distance (3.086 Å), although the metal ions are the same size (6-coordinate radius for Ce^{3+} and U^{3+} = 1.165 Å). The shorter P–U distance has been interpreted as evidence for U–P π-backbonding.

Complexes with monoanionic R_2P^- (phosphido) and di-anionic RP^{2-} (phosphinidene) ligands can be prepared by protonolysis of organometallic precursors as shown in Figure 4.40. The Lu phosphinidene complex acts as a phospha-Wittig reagent and reacts with aldehydes and ketones to form phosphaalkenes.

4.12 **Summary**

Both lanthanoids and actinoids are generally considered as hard Lewis acids and so their coordination chemistry is dominated by hard O- and N-donor ligands. Bonding in most cases is highly ionic, resulting in labile complexes that can have highly irregular coordination geometries dictated by ligand steric factors. Large ionic radii result in high coordination numbers, except where ligands are exceptionally sterically demanding. In aqueous solution, complex formation is usually entropy driven, and thermodynamically stable complexes are only obtained with multidentate ligands such as EDTA^{4-}. The coordination chemistry of the lanthanoids is dominated by the +3 oxidation state, though Ln^{2+} complexes are accessible, particularly for Sm, Eu, and Yb in non-aqueous solution. Ce^{4+} complexes can be stabilized by the correct choice of ligands. The early actinoids display a range of oxidation states in their complexes, but later An have coordination chemistry that closely resembles that of Ln^{3+}. There is evidence of some covalent contributions to bonding in complexes of the early An, which are impossible for corresponding complexes of the Ln: these differences in bonding can be exploited in An/Ln separations that are necessary in reprocessing of nuclear fuels.

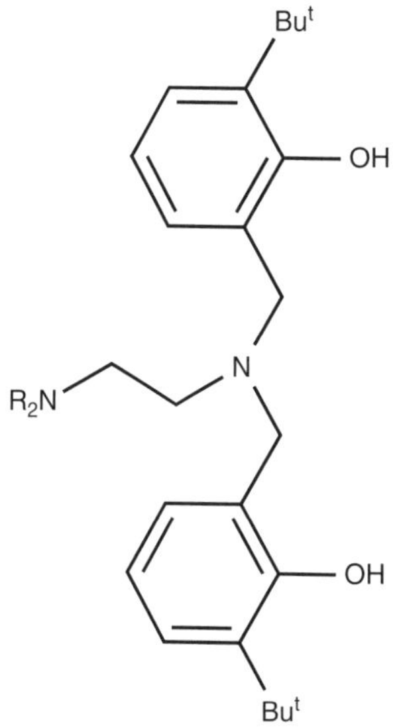

4.13 Exercises

1. Consider the formation of 1:1 complexes in aqueous solution between Ln^{3+} and each of the ligands shown in Figure 4.41. Put the complexes in order of *increasing* stability constant and explain your reasoning.

2. Zhou et al. (*Inorg. Chem.* 2007, **46**: 958) have reported complexes of Yb^{2+} with ligand L. The complexes are formed by reaction of H_2L (see margin) with 1 equiv of $[Yb\{N(SiMe_3)_2\}_2(thf)_2]$.

   (a) Suggest starting materials and conditions for the synthesis of $[Yb\{N(SiMe_3)_2\}_2(thf)_2]$.

   (b) Identify the features of ligand L that make it suitable for the preparation of a monomeric complex of Yb^{2+}.

   (c) Explain how H_2L reacts with 1 equiv of $[Yb\{N(SiMe_3)_2\}_2(thf)_2]$ to form $[Yb(L)(thf)_2]$.

   (d) When 1 equiv of phenylacetylene (PhC≡CH) is added to a solution of $[Yb(L)(thf)_2]$, a colourless gas is evolved and colourless crystals of $[Yb(L)(C_2Ph)]$ are formed. Identify the colourless gas and explain the reaction that has taken place.

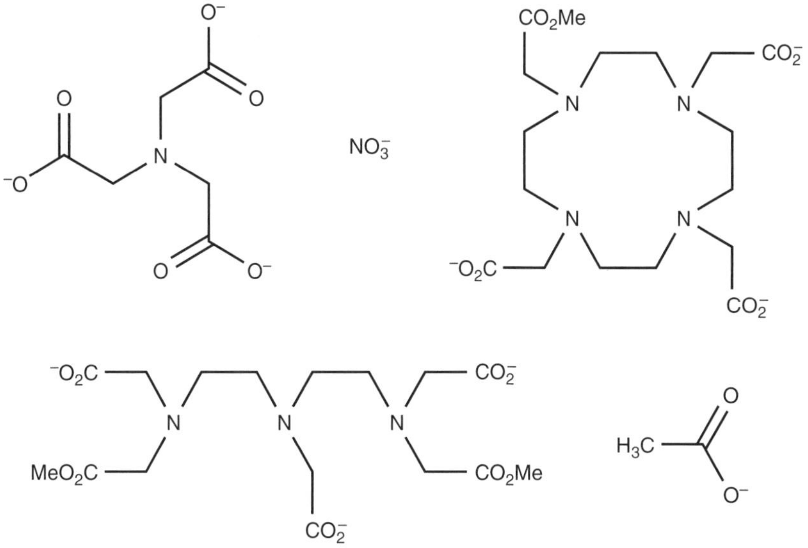

H_2L

	approximate pK_a
Typical phenol	16–18
$HN(SiMe_3)_2$	26
PhC≡CH	29

Figure 4.41 Ligands for Exercise 1.

$(NN'_3)^{3-}$

$[U(NN'_3)]$

3. $[U(NN'_3)]$ is paramagnetic with $\mu_{eff} = 3.06\ \mu_B$.

$[U(NN'_3)]$ reacts with one equiv. of trimethylsilylazide Me_3SiN_3. A colourless gas is evolved, and a new complex **X** is formed. **X** has $\mu_{eff} = 2.17\ \mu_B$.

$[U(NN'_3)]$ reacts with one equiv. of pyridine C_5H_5N to form a new complex **Y**. **Y** has $\mu_{eff} = 2.95\ \mu_B$.

Explain these observations and suggest, with reasons, structures for **X** and **Y**.

Would you expect the reactivity of $[Nd(NN'_3)]$ towards (i) Me_3SiN_3 and (ii) pyridine to be similar to that of $[U(NN'_3)]$? Explain your reasoning.

4.14 Further reading

1. Denning, R.G., 'Electronic structure and bonding in actinyl ions and their analogs'. *Journal of Physical Chemistry A*, 2007, **111**(20): 4125–43.
2. Neidig, M.L., D.L. Clark, and R.L. Martin, 'Covalency in f-element complexes'. *Coordination Chemistry Reviews*, 2013, **257**(2): 394–406.
3. Liddle, S.T., 'The renaissance of non-aqueous uranium chemistry'. *Angewandte Chemie-International Edition*, 2015, **54**(30): 8604–41.
4. Ephritikhine, M., 'Molecular actinide compounds with soft chalcogen ligands'. *Coordination Chemistry Reviews*, 2016, **319**: 35–62.
5. Goodwin, C.A.P. and D.P. Mills, 'Silylamides: towards a half-century of stabilising remarkable f-element chemistry', in *Organometallic Chemistry*, 2017, **41**: 123–56.
6. Nielsen, L.G., A.K.R. Junker, and T.J. Sorensen, 'Composed in the f-block: solution structure and function of kinetically inert lanthanide(iii) complexes'. *Dalton Transactions*, 2018, **47**(31): 10360–76.

Organometallic chemistry

5.1 Introduction

The organometallic chemistry of the f-elements is very different from that of the d-transition metals: this is mainly due to the predominantly ionic nature of the bonding and the lack of two-electron redox chemistry for the f-elements. This chapter will open with examples of homoleptic σ-bonded alkyls and aryls and then move on to homoleptic π-bonded complexes, some of which display 'non-classical' oxidation states (e.g. Ln(0) and Ln(II)). The π-bonded organometallics have been the subjects of numerous experimental and theoretical investigations aimed at understanding the bonding, and particularly the possibility of f-orbital contributions.

Some of the unique reactions of σ-bonded organometallics will be discussed for heteroleptic complexes in which the reactivity of the σ-bonded alkyl/aryl is moderated by the presence of stabilizing ancillary ligands. Finally, some examples of organo-f-element catalysed reactions will be presented.

A homoleptic complex is one in which all the ligands are the same. A heteroleptic complex is one in which the ligands are not all the same.

5.2 Homoleptic σ-bonded alkyl and aryl complexes

Alkyl R$^-$ ligands are soft Lewis bases and powerful Brønsted bases and their complexes with the hard Lewis acid Ln and An ions are predominantly ionic in nature and therefore highly reactive. The examples chosen here show how the large Ln and An ions require ligands that are either chelating or sterically very demanding in order to stabilize monomeric neutral complexes. Alternatively, anionic 'ate' complexes with higher coordination numbers can be prepared.

5.2.1 Homoleptic lanthanoid complexes

Synthetic routes to some representative homoleptic σ-bonded organometallics of the lanthanoids are shown in Figure 5.1. The syntheses shown here all make use of salt exchange reactions under strictly anaerobic and anhydrous conditions. These examples illustrate how monomeric species can be obtained by formation of anionic 'ate' complexes ([Li(tmeda)]$_3$[LnMe$_6$] and [Li(tmeda)$_2$]

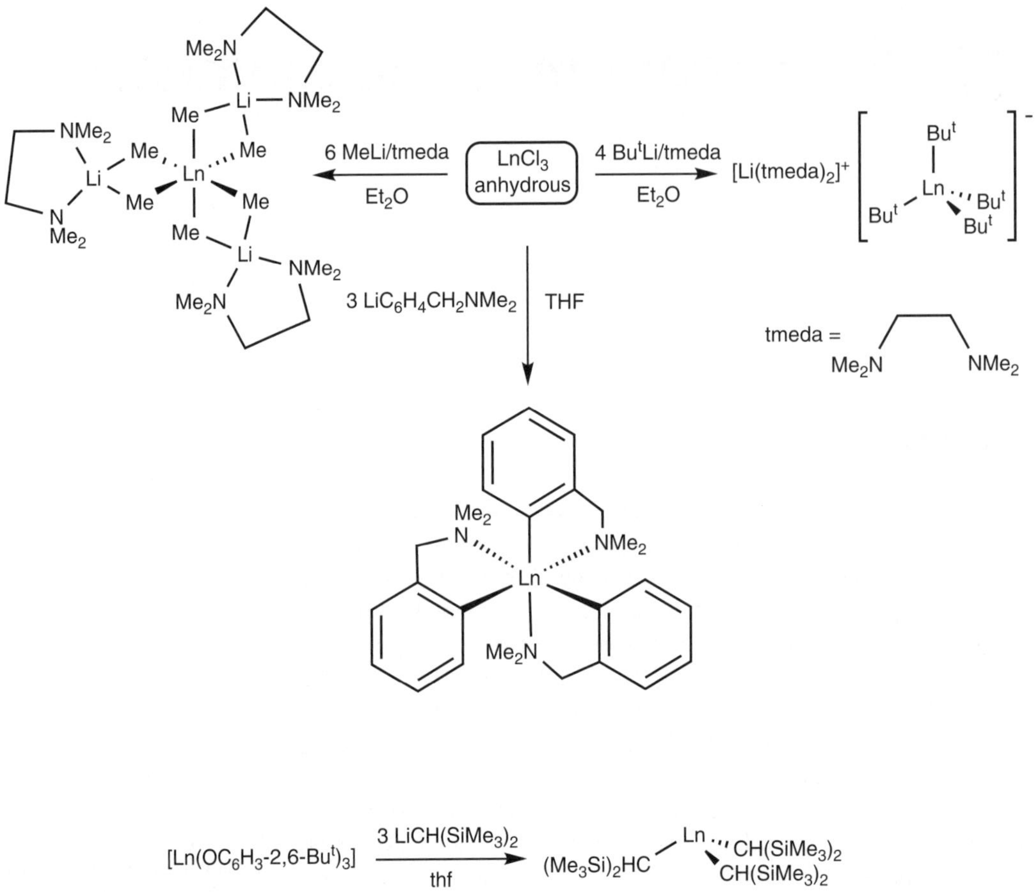

Figure 5.1 Synthesis of homoleptic σ-bonded lanthanoid alkyls

[LnBu$^t_4$]). Use of the extremely bulky CH(SiMe$_3$)$_2^-$ ligand allows formation of a three-coordinate complex [Ln{CH(SiMe$_3$)$_2$}$_3$] that has a trigonal pyramidal structure analogous to that of the lanthanoid tris-silylamide complexes [Ln{N(SiMe$_3$)$_2$}$_3$] (see Chapter 4).

Homoleptic bis alkyl Ln(II) complexes [Ln{C(SiMe$_3$)$_3$}$_2$] can be prepared by salt exchange reaction of LnI$_2$ with KC(SiMe$_3$)$_3$. The Ln^{2+} ions are larger than their Ln^{3+} counterparts and so an extremely bulky ligand is required to form a monomeric complex.

All of the [Ln{C(SiMe$_3$)$_3$}$_2$] have C(alkyl)–Ln–C(alkyl) angles of approximately 136°–137°, very similar to the Cp(centroid)–Ln–Cp(centroid) angles in the corresponding [Cp*$_2$Ln] (Cp* = η^5-C$_5$Me$_5$) complexes (see Section 5.3.2). The similarity of the C–Ln–C angles for Ln = Eu or Yb suggests that the origin of the bending is electronic rather than steric, and it is suggested that there is metal d-orbital involvement in the bonding. The bending is accompanied by Ln–Me agostic interactions, three for Ln = Eu and two for Ln = Yb. The structure of [Yb{C(SiMe$_3$)$_3$}$_2$] is shown in Figure 5.2.

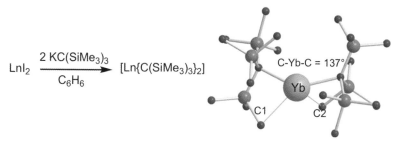

Figure 5.2 Synthesis and structure of [Yb{C(SiMe₃)₃}₂]. Yb–C1 = 2.85 Å, Yb–C2 = 2.90 Å

5.2.2 Homoleptic actinoid complexes

Early attempts to synthesize homoleptic actinoid alkyls were plagued with diffi-culties including thermal instability, and reduction of $AnCl_4$ starting materials to metallic An, but a small number of complexes have now been characterized. As expected for early actinoids, a variety of oxidation states are available, and rep-resentative examples of U complexes are shown in Figure 5.3.

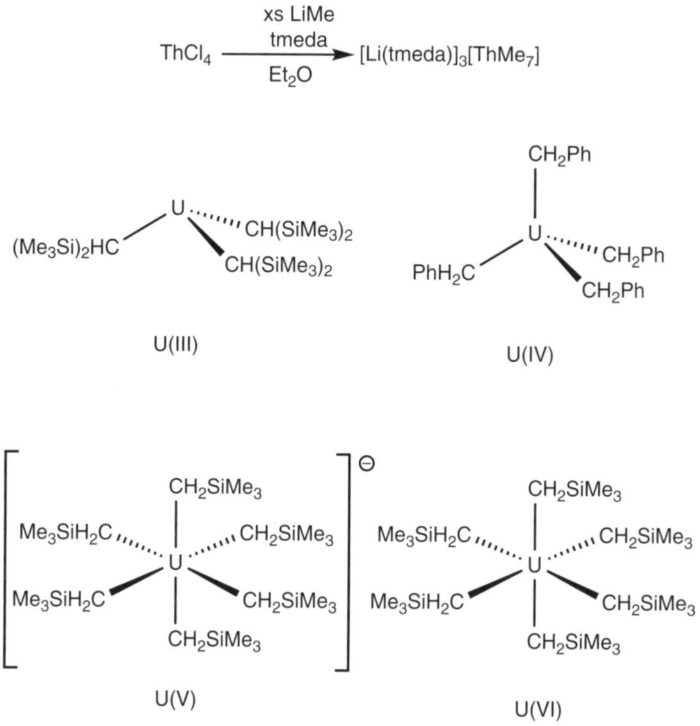

$$ThCl_4 \xrightarrow[\text{Et}_2\text{O}]{\overset{\text{xs LiMe}}{\text{tmeda}}} [Li(tmeda)]_3[ThMe_7]$$

U(III)

U(IV)

U(V)

U(VI)

Figure 5.3 Homoleptic σ-bonded actinoid alkyls

5.3 Homoleptic cyclopentadienyl complexes

Cyclopentadienyl (Cp; η^5-$C_5H_5^-$) is ubiquitous in organometallic chemistry. The first d-transition metal Cp complex, ferrocene (Cp_2Fe), was characterized in 1952, and examples of lanthanoid complexes followed soon after in 1954. Cp can be tuned sterically and electronically by introducing substituents onto the ring and it is widely used as an ancillary or 'spectator' ligand in organometallic chemistry. Cp complexes of f-elements are generally less moisture sensitive than alkyl complexes due to the reduced Brønsted basicity of Cp^- (pK_a for HCp = 18 cf 56 for CH_4). Cp complexes have been the subjects of many structural, spectroscopic, and theoretical studies aimed at understanding the bonding.

5.3.1 Homoleptic Ln(III) complexes

[Cp_3Ln] have been prepared for all the lanthanoids apart from Pm, and they have been characterized by x-ray crystallography. The complexes are synthesized by salt exchange reaction under anhydrous and anaerobic conditions to form initially the thf adduct [$Cp_3Ln(thf)$], which can be desolvated *in vacuo* to give the solvent-free product.

The structures of Cp_3Ln vary as Ln^{3+} radius decreases along the lanthanoid series as shown in Figure 5.4. There are three distinct structural types in the solid state:

(i) Cp_3La adopts a polymeric structure with three η^5-Cp ligands and one bridging η^2-Cp

(ii) in the middle of the series, a monomeric structure with three η^5-Cp ligands is adopted

$$LnCl_3 + 3NaCp \xrightarrow[\text{anhydrous}]{\text{thf}} Cp_3Ln(thf) + 3NaCl_{ppt} \xrightarrow[\text{vacuum}]{200°C} Cp_3Ln$$

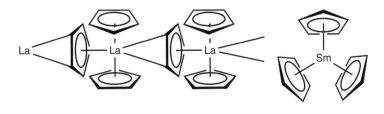

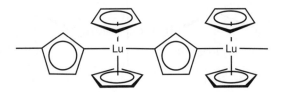

Figure 5.4 Synthesis and structures of Cp_3Ln

(iii) at the end of the series, Lu^{3+} is too small to accommodate three η^5-Cp ligands, and so Cp_3Lu is polymeric with two η^5-Cp ligands and one μ_2-Cp.

The bonding in Cp_3Ln is electrostatic and the complexes are labile as show by the facile exchange reaction with $FeCl_2$ to form ferrocene as shown in Figure 5.5.

5.3.2 Homoleptic bis(cyclopentadienyl) Ln(II) complexes: Eu, Yb, Sm

Eu and Yb have the most accessible +2 oxidation states, and these were the first lanthanoids for which Cp_2Ln were prepared. Metallic Eu and Yb are both soluble in liquid NH_3 and complexes were prepared as shown in Figure 5.6.

Ln^{2+} ions have larger radii than their Ln^{3+} analogues and so solvent-free Cp_2Ln are very coordinatively unsaturated and form polymers in the solid state.

In order to form molecular species the more sterically demanding Cp* (C_5Me_5) ligand is required, and $[Cp^*{}_2Ln]$ (Ln = Eu, Yb, or Sm) can be prepared by a salt exchange reaction from LnI_2 as shown in Figure 5.7.

$$2Cp_3Ln + 3FeCl_2 \longrightarrow 2LnCl_3 + 3Cp_2Fe$$

Figure 5.5 Reaction of Cp_3Ln with $FeCl_2$

$$Ln + 2HCp \xrightarrow[NH_3]{\text{inert atmosphere}} [Cp_2Ln(NH_3)] + H_2 \xrightarrow[\text{vacuum}]{\Delta H} \begin{array}{c} Cp_2Ln \\ \text{polymeric} \end{array}$$

Figure 5.6 Synthesis of Ln(II) bis(cyclopentadienyls)

$$\begin{array}{c} LnI_2 + 2NaCp^* \\ \text{anhydrous} \end{array} \xrightarrow[\text{thf}]{\text{inert atmosphere}} Cp^*{}_2Ln(thf)_2 + 2NaI \xrightarrow[\text{vacuum}]{\Delta H} \begin{array}{c} Cp^*{}_2Ln \\ \text{monomeric} \end{array}$$

$$Cp^* = C_5Me_5$$

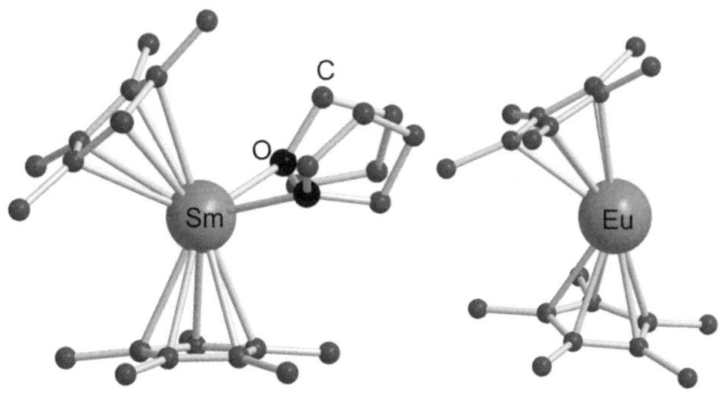

Figure 5.7 Synthesis and structures of $[Cp^*{}_2Ln(thf)_2]$ (left) and $[Cp^*{}_2Eu]$ (right). $Cp^*{}_{centroid}$–Ln–$Cp^*{}_{centroid} = 136.7$ ($[Cp^*{}_2Sm(thf)_2]$) $Cp^*{}_{centroid}$–Ln–$Cp^*{}_{centroid} = 140°$ ($[Cp^*{}_2Eu]$)

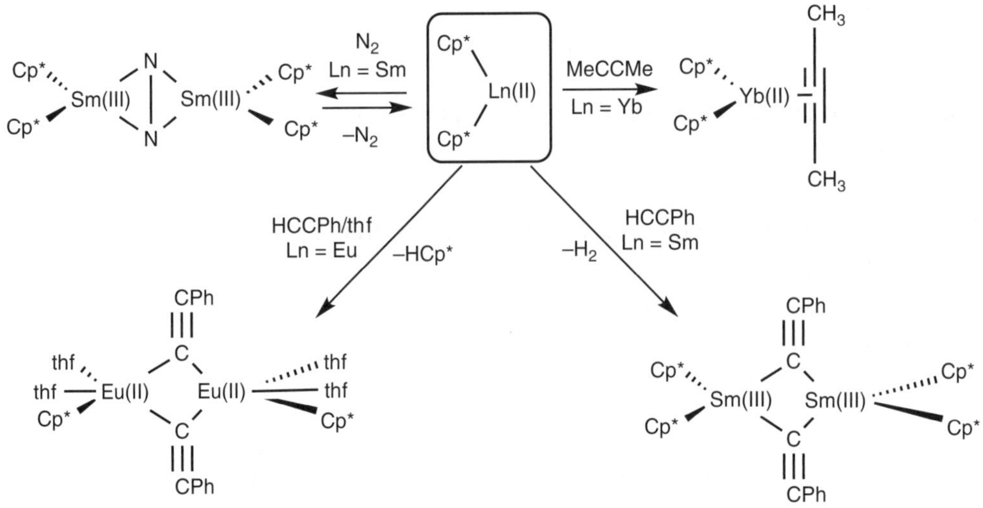

Figure 5.8 Reactions of $[Cp^*_2Ln]$ ($Cp^* = \eta^5\text{-}C_5Me_5$)

Both $[Cp^*_2Ln(thf)_2]$ and $[Cp^*_2Ln]$ have been characterized by x-ray diffraction (see Figure 5.7). Not surprisingly, the Cp^* rings of $[Cp^*_2Ln(thf)_2]$ are tilted away from a parallel arrangement in order to accommodate equatorial thf ligands. However, it is found that the solvent-free $[Cp^*_2Ln]$ also adopt the 'bent metallocene' structure with a Cp^*–Ln–Cp^* angle almost identical to that in the thf adduct. The 'bent metallocene' structure is also observed for Cp^*_2M (M = Ca, Sr, Ba), but contrasts with the parallel Cp rings found in d-transition metal analogues such as ferrocene.

Reactivity of solvent-free $[Cp^*_2Ln]$ is dominated by two factors: (i) the complexes are coordinatively unsaturated due to the large Ln^{2+} radii (e.g. 6-coordinate radius for $Eu^{2+} = 1.31$ Å cf $Eu^{3+} = 1.09$ Å) and (ii) Ln^{2+} readily undergo 1-electron oxidation as illustrated in Figure 5.8. Adduct formation with Lewis bases including phosphines and electron-rich alkenes and alkynes is also observed. $[Cp^*_2Sm]$ reacts reversibly with N_2 to form a Sm(III) complex of N_2^{2-} in which the N-N distance is consistent with a N=N double bond. Sm(II) is a more powerful reducing agent than Eu(III) and this is illustrated by the reactions of $[Cp^*_2Sm]$ and $[Cp^*_2Eu]$ with phenylacetylene. PhCCH oxidizes Sm(II) whereas it acts as an acid towards $[Cp^*_2Eu]$.

5.3.3 Homoleptic tris(cyclopentadienyl) Ln(II) complexes

Eu, Yb, and Sm have the most accessible +2 oxidation states, but since the1960s other Ln(II) have been known in the solid state: e.g. dihalides $[Ln^{2+}][X^-]_2$ of Tm, Dy, Nd can be prepared by reduction of LnX_3 with metallic Ln. These lanthanoids were therefore targets in the search for new Ln(II) cyclopentadienyls. The reduction of $[Cp'_3Ln]$ complexes with either metallic K or KC_8 (potassium graphite) in the presence of a ligand to bind K^+ (either 18-crown-6 or crypt(2.2.2), as shown in Figure 5.9, formed isolable $[KL]^+[Cp'_3Ln]^-$ salts. This reaction has now been applied to the whole lanthanoid series (except Pm) and to Y.

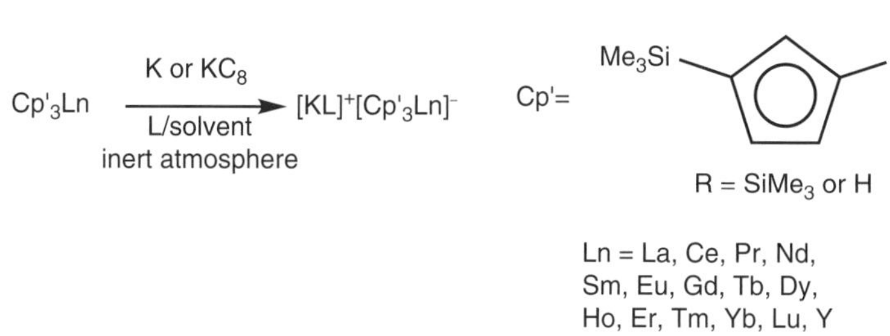

$$Cp'_3Ln \xrightarrow[\substack{L/solvent \\ inert\ atmosphere}]{K\ or\ KC_8} [KL]^+[Cp'_3Ln]^-$$

Cp'=

R = SiMe₃ or H

Ln = La, Ce, Pr, Nd,
Sm, Eu, Gd, Tb, Dy,
Ho, Er, Tm, Yb, Lu, Y

Figure 5.9 Synthesis of $[Cp_3Ln]^-$

Table 5.1 Structural data and electron configurations for $[Cp'_3Ln]^-$

Ln	Δ(Ln–Cp')*/pm	Electron config. of Ln^{2+}
Sm, Eu, Tm, Yb	12.3 to 15.6	$4f^{n+1}$
La, Ce, Pr, Nd, Gd, Dy, Ho, Er, Lu	2.7 to 3.2	$4f^n5d^1$ n = 0 (La) to 14 (Lu)

*Difference in Ln–Cp'$_{cent}$ distance between $[Cp'_3Ln]^-$ and $[Cp'_3Ln]$

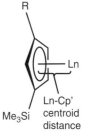

Structural and spectroscopic studies show that $[KL]^+[Cp'_3Ln]^-$ fall into two distinct classes as shown in Table 5.1.

The four Ln that have the most accessible +2 oxidation states (Sm, Eu, Tm, and Yb) have a large difference in Ln–Cp' centroid distance between $[Cp'_3Ln]$ and $[Cp'_3Ln]^-$. This is consistent with formation of a 'classical' Ln^{2+} ion with the extra electron entering a 4f orbital on reduction of the Ln(III) precursor. The other Ln show very small differences in Ln–Cp' distances between Ln(III) and Ln(II) complexes, consistent with a $4f^n5d^1$ configuration for the Ln(II) species.

This observation is analogous to the structures of Ln diiodides LnI_2. For Ln = Sm, Eu, Tm, Yb the diiodides are salt-like $4f^n[Ln^{2+}][I^-]_2$ whereas for Ln = La, Ce, Pr, Nd, Gd, Dy, Ho, Er, Lu they are $[Ln^{3+}][I^-]_2[e^-]_2$ with the additional electrons accommodated in delocalized 5d bands (see Chapter 3). Electronic absorption spectra of the $[Cp_3Ln]^-$ species support the proposed electron configurations. $[Cp_3Ln]$ species have been the subject of numerous theoretical studies, and it has been shown that the C_{3v} ligand field due to three η^5-Cp rings lowers the energy of the Ln $5d(z^2)$ orbital to the extent that it is comparable to that of the 4f orbitals.

5.3.4 Homoleptic cyclopentadienyl complexes of the actinoids

Homoleptic cyclopentadienyl complexes of the actinoids are known in oxidation states from +4 to +2.

The +4 oxidation state is readily accessible for early actinoids and $[Cp_4An]$ (An = Th, Pa, U, Np) can be prepared by a salt exchange reaction as shown in Figure 5.10.

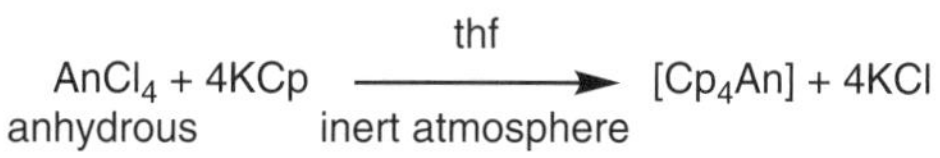

$$\underset{anhydrous}{AnCl_4 + 4KCp} \xrightarrow[\substack{inert\ atmosphere}]{thf} [Cp_4An] + 4KCl$$

Figure 5.10 Synthesis of $[Cp_4An]$

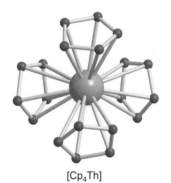

[Cp$_4$Th]

In the solid state [Cp$_4$An] have a pseudotetrahedral structure with S$_4$ symmetry, but in solution they are tetrahedral with four equivalent Cp ligands. [Cp$_4$Th] (5f^0) is diamagnetic and its ^{1}H NMR spectrum shows a singlet at δ 6.3. [Cp$_4$U] (5f^2) is isostructural and so, due to its tetrahedral symmetry in solution, would be expected to exhibit no pseudocontact shifting in its NMR spectrum (see Chapter 2). It is therefore significant that the ^{1}H NMR spectrum of [Cp$_4$U] shows a singlet at δ 13, very different from that of the Th analogue. This shifting has been interpreted as evidence for delocalization of unpaired f-electron density onto the Cp ligands.

Monomeric [Cp″$_3$An] (Cp″ = substituted Cp, e.g. C$_5$Me$_5$, C$_5$H$_4$SiMe$_3$ etc) have been known for U and Th (for which the +3 oxidation state is extremely rare) since 1986, and more recently for other An. These complexes have been the subjects of numerous spectroscopic, theoretical, and structural studies, many with the aim of understanding the nature of bonding. Their structures are analogous to those of corresponding lanthanoid complexes, with the three cyclopentadienyl centroids coplanar with the central metal atom.

[Cp″$_3$An]$^-$ complexes with Th and U in the unprecedented +2 oxidation state were reported in 2015 and 2013 respectively. The synthesis used the same reduction reaction with K or KC$_8$ that was used for the lanthanoid(II) analogues, and the Ln and An complexes are isostructural.

5.4 Bonding in [Cp$_3$Ln] and [Cp$_3$An]

Monomeric [Cp$_3$Ln] and [Cp$_3$An] have C$_{3v}$ symmetry with the three Cp centroids coplanar with the metal atom, and in this point group the π$_1$ HOMOs of the three Cp$^-$ ligands transform as a$_1$ + a$_2$ + 2e. The π$_1$-derived levels are split strongly by inter-ligand interactions, and the a$_2$ level is significantly destabilized. In C$_{3v}$ symmetry, the metal d-orbitals transform as a$_1$ + 2e and so there is no metal d-orbital that has the correct symmetry to interact with the a$_2$ ligand combination. However, f(y(3x^2-y^2)) transforms as a$_2$, and so is capable of stabilizing the a$_2$ ligand combination in addition to contributing significantly to metal-ligand bonding. The other f-orbitals (which transform as 2a$_1$ + 2e) are essentially non-interacting.

A simplified MO scheme for [Cp$_3$M] is shown in Figure 5.11.

Theoretical studies show that as the actinoid series is traversed, the energy of the 5f orbitals decreases and there is an increase in 5f orbital contributions to metal-ligand bonding, along with a decrease in 6d orbital contribution to bonding. However, these changes are not accompanied by an increase in electron density in the internuclear region (what chemists usually mean by 'covalency'). Using the measure of internuclear electron density, bonding in [Cp$_3$Ln] is essentially ionic whereas that for early [Cp$_3$An] has some covalent contribution, which decreases as the actinoid series is traversed.

[Cp$_3$Ln] have also been investigated extensively, and a combination of spectroscopic and theoretical studies indicates that in [Cp$_3$Yb] there is a very significant component of Cp$^-$ to 4f charge transfer in the bonding. This effect is predicted for other easily reduced Ln^{3+} (e.g. Sm and Eu), but there is unlikely to be 4f contribution to bonding in other [Cp$_3$Ln].

f(y(3x^2−y^2)) viewed along z-axis

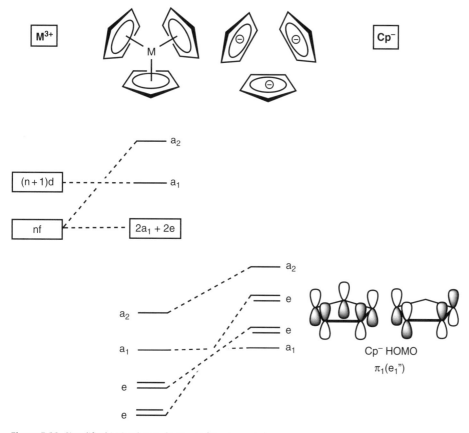

Figure 5.11 Simplified MO scheme for Cp₃M (M = Ln or An)

5.5 Comparison of Lewis acidity of [Cp₃Ln] and [Cp₃An]

Both [Cp₃Ln] and [Cp₃An] species are Lewis acidic and can form adducts with a range of Lewis bases, including some examples of π-acid ligands such as isocyanides (RNC) (Table 5.2).

The increase in ν_{CN} on coordination of CNC_6H_{11} to [Cp₃Pr] indicates that the ligand is acting as a σ-donor, with no π-backbonding, whereas the decrease in ν_{CN} on complexation of CNC_6H_4OMe to $[(C_5Me_4H)_3U]$ is strong experimental evidence for a significant contribution of π-backbonding from U 5f orbitals into ligand π^* orbitals in addition to σ-donation from ligand to metal 6d(z^2). $[(C_5Me_4H)_3U]$ forms an isolable adduct with CO in which ν_{CO} has decreased by 170 cm^{-1} from that of free CO, consistent with U-to-CO π-backbonding. No CO adduct can be isolated for [Cp₃Pr] due to the inability of the lanthanoid 4f orbitals to take part in covalent interactions with ligands.

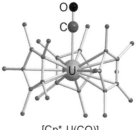

[Cp*₃U(CO)]

8 π electrons
non-planar, antiaromatic

↓ 2K

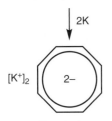

$[K^+]_2$

10 π electrons
planar, aromatic

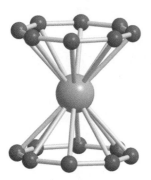

Uranocene
$[U(\eta^8-C_8H_8)_2]$

Table 5.2 IR data for adducts of $[Cp_3Pr]$ and $[Cp'_3U]$ with π-acid ligands ($Cp' = C_5Me_4H$)

	$[Cp_3Pr(CNC_6H_{11})]$	$[Cp'_3U(CNC_6H_4OMe)]$	$[Cp'_3U(CO)]$
$\nu_{CN/CO}$(complex)	2178 cm^{-1}	2072 cm^{-1}	1880 cm^{-1}
$\nu_{CN/CO}$(free ligand)	2130 cm^{-1}	2122 cm^{-1}	2143 cm^{-1}

5.6 Cyclooctatetraenyl (COT^{2-}) complexes

Cyclooctatetraene C_8H_8 is an 8 π electron antiaromatic molecule that can readily be reduced to give the planar aromatic 10 π electron $C_8H_8^{2-}$ ion. In 1963, it was recognized that the nodal properties of f-orbitals have the correct symmetry to overlap with the π orbitals of $C_8H_8^{2-}$ and the existence of uranocene $[U(C_8H_8)_2]$, an f-element analogue of ferrocene $[Cp_2Fe]$, was predicted. The prediction became reality in 1968 with the synthesis and crystallographic characterization of uranocene, and the Pa and Np analogues followed shortly afterwards. Uranocene has D_{8h} symmetry with parallel $C_8H_8^{2-}$ rings in an eclipsed conformation.

The ligand orbitals of most interest are the π_2 HOMOs, which transform as e_{2g} and e_{2u}, and the π_3 LUMOs, which transform as e_{3g} and e_{3u} in $[U(COT)_2]$ (Figure 5.12). The 5f orbitals are split by the D_{8h} ligand field into e_{3u}, a_{2u}, e_{1u}, e_{2u}, and the 6d orbitals (which interact more strongly with the ligand field due to their greater radial extent) are split into a_{1g}, e_{2g}, e_{1g}. This means that a δ-interaction between occupied ligand orbitals and unoccupied U 5f and 6d orbitals is symmetry allowed. A φ-interaction between the occupied uranium e_{3u} orbitals and the ligand e_{3u} LUMO is also symmetry allowed. Uranocene (U^{4+}, $5f^2$) has a triplet ground state that has been confirmed experimentally. As predicted by this scheme, $[Th(C_8H_8)_2]$ (Th^{4+}, $5f^0$) is diamagnetic, as is $[Pu(C_8H_8)_2]$ (Pu^{4+}, $5f^4$). $[Np(C_8H_8)_2]$ ($4f^3$) has a doublet ground state. There is significant covalent contribution to the bonding in actinocenes, with a greater contribution from 6d than 5f orbitals.

Cerocene, $[Ce(C_8H_8)_2]$, is prepared from $[Ce(O^iPr)_4(^iPrOH)]_2$ by reaction with $AlEt_3$ and C_8H_8. It might naively be considered as a lanthanoid analogue of the An(IV) actinocenes, with which it is isostructural. However, Ce(IV) is a powerful oxidizing agent and so is expected to be incompatible with $C_8H_8^{2-}$ ligands. Cerocene has been the subject of numerous theoretical and spectroscopic studies. It has a $^1A_{1g}$ ground state which is best described as Ce^{3+} ($4f^1$) with two $(C_8H_8)^{1.5-}$ ligands and antiferromagnetic coupling.

Theoretical studies show that $[An(C_8H_8)_2]$ complexes of the later actinoids for which the +3 oxidation state is the most important, have electronic structures analogous to that of cerocene.

A series of anionic Ln(III) complexes $[Ln(C_8H_8)_2]^-$ is also known; there is no evidence of significant covalent contributions to their bonding.

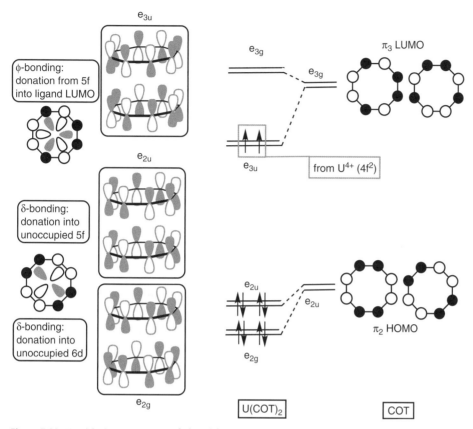

Figure 5.12 Simplified MO scheme for [U(COT)$_2$]

5.7 **Complexes with neutral arenes**

Complexes of d-transition metals with neutral arene ligands are very well known (e.g. [Cr(η^6-C$_6$H$_6$)$_2$]), and in these complexes there is a significant component of backbonding from filled metal d-orbitals into the benzene LUMO. The reaction scheme in Figure 5.13 (analogous to synthesis of [Cr(η^6-C$_6$H$_6$)$_2$]) forms a U(III) complex with a neutral arene ligand.

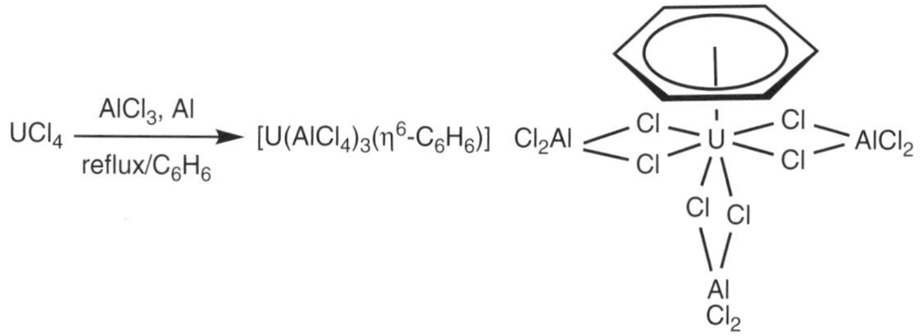

$$UCl_4 \xrightarrow[\text{reflux/C}_6\text{H}_6]{\text{AlCl}_3, \text{Al}} [U(AlCl_4)_3(\eta^6\text{-C}_6\text{H}_6)]$$

Figure 5.13 Synthesis of [U(AlCl$_4$)$_3$(η^6-C$_6$H$_6$)]

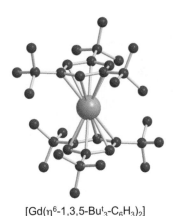

[Gd(η^6-1,3,5-But$_3$-C$_6$H$_3$)$_2$]

5.7.1 Bis(arene) lanthanoid(0) complexes

Synthesis of an f-element analogue of [Cr(η^6-C$_6$H$_6$)$_2$] was finally achieved in 1987 when [Gd(η^6-1,3,5-But$_3$-C$_6$H$_3$)$_2$] was first reported. The synthesis was achieved by reacting metal vapour with 1,3,5-But$_3$-C$_6$H$_3$. The synthesis was successfully repeated for other lanthanoids, and a series of complexes for Ln = Nd, Gd, Tb, Dy, Ho, and Er has been prepared.

[Ln(η^6-1,3,5-But$_3$-C$_6$H$_3$)$_2$] have a sandwich structure with parallel neutral arene ligands that adopt a staggered conformation. The extremely bulky arene ligand is required in order to provide some steric protection for the large Ln(0) atom, and the staggered conformation is adopted to reduce inter-ligand steric repulsions between But substituents.

The bonding in [Ln(η^6-1,3,5-But$_3$-C$_6$H$_3$)$_2$] has been analysed by analogy with that in the 18-electron complex [Cr(η^6-C$_6$H$_6$)$_2$], using spectroscopic and computational studies.

There are two components to the arene–Ln bonding: ligand-to-metal donation (there are metal valence orbitals of the correct symmetry to interact with ligand a_{1g}, a_{2u}, e_{1g}, e_{1u} orbitals), and metal-to-ligand back donation (metal d(x^2-y^2) and d(xy) orbitals can interact with the ligand e_{2g} LUMO combination) as shown in Figure 5.14.

This bonding scheme requires the metal to have a d^{1}s^2 electron configuration, and as we saw in Chapter 1, most of the Ln atoms have a 4f^{n}5d^{0}6s^2 configuration. The stability of [Ln(η^6-1,3,5-But$_3$-C$_6$H$_3$)$_2$] can be correlated with

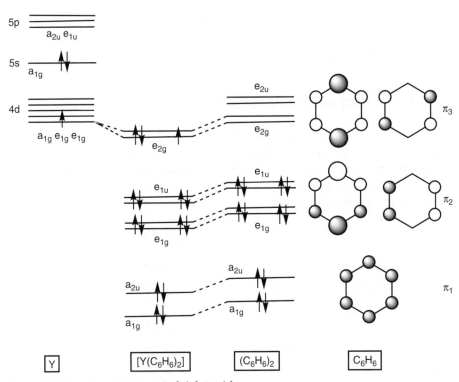

Figure 5.14 Simplified MO scheme for [Y(η^6-C$_6$H$_6$)$_2$]

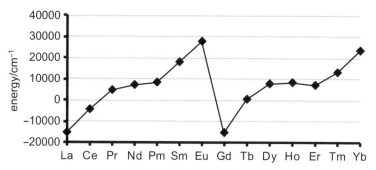

Figure 5.15 $4f^n$ to $4f^{n-1}5d^1$ promotion energies

$4f^n{\rightarrow}4f^{n-1}5d^1$ promotion energies (see Figure 5.15) and the complexes can be isolated for Ln = Nd, Gd, Tb, Dy, Ho, and Er. Although the $4f^n{\rightarrow}4f^{n-1}5d^1$ promotion energies appear to be favourable for La, Ce, and Pr, the radii of these early Ln are too large for the complexes to be stable. The Ln–arene interaction is by no means weak: the bond dissociation enthalpy is 301 kJ mol^{-1} for Y and 197 kJ mol^{-1} for Dy.

5.8 Heteroleptic organometallic complexes

Homoleptic σ-bonded alkyls of the lanthanoids and actinoids are highly reactive, often in an uncontrollable way. In order to moderate the reactivity of the metal–alkyl bond, ancillary ligands that are less reactive than the alkyl ligand are introduced to form heteroleptic complexes that have more controlled reactivity. The ancillary ligands can be chosen to tune the reactivity of the complexes by steric or electronic effects. Ancillary ligands should fulfil some or all of the following:

- Be less reactive than σ-bonded alkyls
- E.g. for anionic ligands L$^-$, pK_a (L–H) < pK_a (R–H) (pK_a (CH$_4$) ≈ 56)
- Increase coordinative saturation at the metal atom
- Have tunable steric/electronic properties
- Stabilize heteroleptic complexes with respect to ligand redistribution reactions.

5.8.1 Ancillary ligands

Cyclopentadienyl η^5-C$_5$H$_5^-$ is probably the most widely used ancillary ligand (Figure 5.16(a)). The pK_a of HCp is ≈ 18 so C$_5$H$_5^-$ is a much weaker Brønsted base than an alkyl ligand and hence much less reactive. Another advantage of Cp$^-$ is that it can be modified by substitution on the ring to modify steric bulk and solubility. Linked or 'ansa' Cp ligands have two effects: they increase

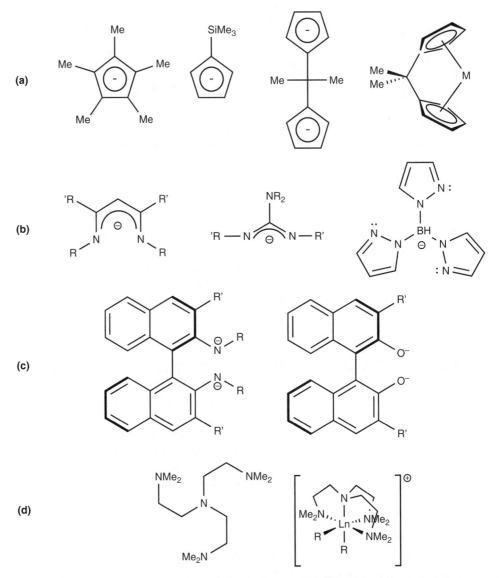

Figure 5.16 Ancillary ligands that have been used in organo-f-element chemistry: (a) Cp and related ligands (b) monoanionic chelating N-donors (c) dianionic chiral binaphthyl derived ligands (d) neutral tetradentate N-donor

stability compared with two separate Cp ligands due to the chelate effect, and they increase accessibility of the remaining alkyl group which is important in catalytic reactions.

Lanthanoid and actinoid ions are hard Lewis acids and so have an affinity for O- and N-donors such as those shown in Figure 5.16 (b), (c), and (d). There is a growing organo-f-element chemistry with these non-Cp ligands.

5.9 Reactivity of σ-bonded alkyls of lanthanoids and actinoids

The reactivity of alkyl f-element complexes has mainly been elucidated by study of heteroleptic complexes, particularly those with Cp-type ancillary ligands. The reactivity of Ln/An-alkyls is dominated by the ionic nature of the metal to carbon bond and by the electrophilic nature of the d^0 metal centres, which result in unique chemistry for these complexes. Due to the lack of 2-electron redox chemistry for the majority of f-elements (U is an exception), the oxidative addition–reductive elimination reactivity that is common in organometallic chemistry of late d-transition metals is not available for organo-f-element complexes.

5.9.1 σ-bond metathesis

σ-bond metathesis is a concerted reaction in which a M–L σ-bond is exchanged with a σ-bond of an incoming substrate as shown in Figure 5.17. The outcome of the reaction depends on the orientation of X and Y in the transition state.

One of the earliest reports of such a reaction was in 1983 when [Cp*$_2$Lu–CH$_3$] was shown to undergo a facile exchange reaction with ^{13}CH$_4$.

Experimental studies showed that $\Delta S^{\ddagger}$ is negative, consistent with an ordered transition state, and that there is a primary H/D kinetic isotope effect, consistent with E–H bond breaking in the rate-determining step.

Computational studies indicate that the transition state is kite-shaped with H$_3$C–H–CH$_3$ angle between 148° and 174°, and that there is charge separation consistent with the highly polar nature of the Ln–C bond.

The reaction shown in Figure 5.18 is not very useful from a synthetic point of view: although C–H activation has been achieved, the outcome of the reaction is

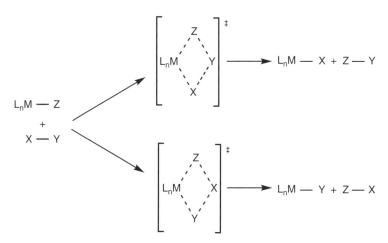

Figure 5.17 σ-bond metathesis

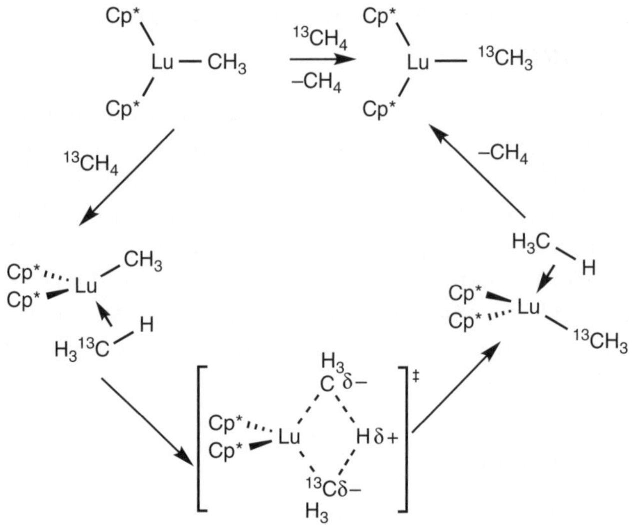

Figure 5.18 C–H activation via σ-bond metathesis

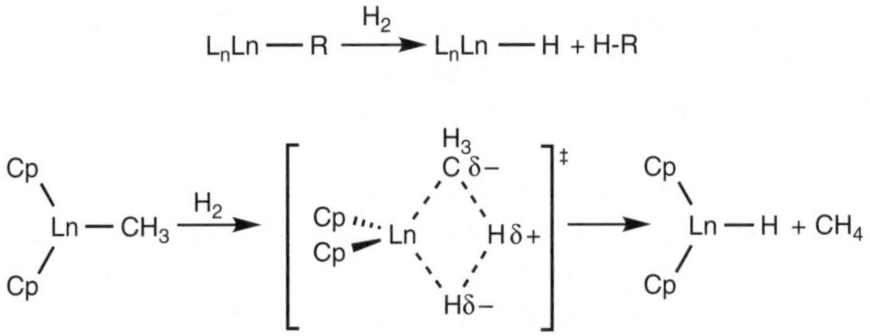

Figure 5.19 Hydrogenolysis of Ln–alkyl via σ-bond metathesis

just exchange of one methyl group for another. The alternative transition state in which the positions of the H and CH_3 are reversed, is energetically unfavourable. σ-bond metathesis of H_2 with alkyl lanthanoid and actinoid complexes forms hydrido complexes that are often highly active catalysts for organic transformations as shown in Figure 5.19.

5.9.2 Alkene insertion

Alkene insertion into a Ln–C bond was first reported in 1978, when Cp'_2LnMe complexes were found to be highly active catalysts for alkene polymerization. The first step of the insertion reaction is π-coordination of the alkene to the metal centre followed by the formation of a 4-centre transition state. The reaction is normally exothermic and irreversible. With unsymmetrical alkenes, the regioselectivity is determined by steric factors: if the ancillary ligands are very sterically demanding, the large L_nLn moiety binds to the least hindered end of the alkene substrate as shown in Figure 5.20.

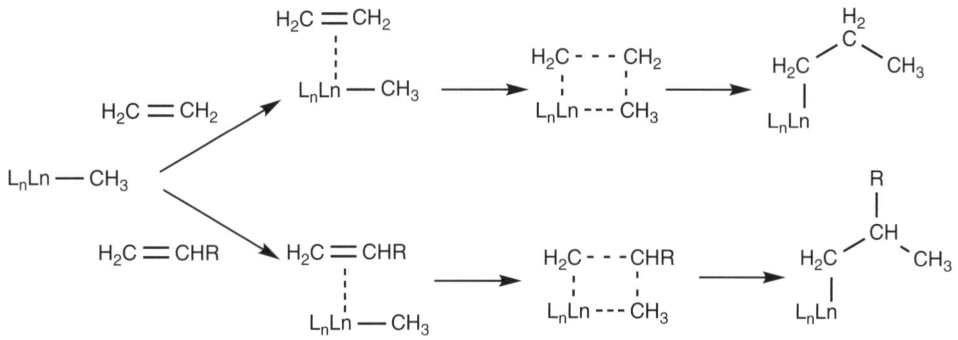

Figure 5.20 Alkene insertion into Ln–alkyl bond

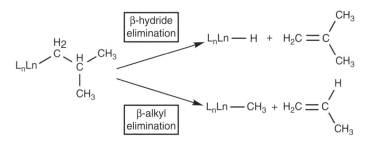

Figure 5.21 Elimination reactions of Ln alkyls

5.9.3 **Elimination reactions**

Elimination reactions are a common route for the degradation of f-element al-kyls. The β-hydride elimination reaction is well-known throughout organome-tallic chemistry; the β-alkyl elimination is less common and was first observed for organo-lutetium complexes in 1983. Both of these elimination reactions are important chain termination steps in organo-f-element catalysed alkene polymerization reactions. The relative prevalence of β-hydride vs. β-alkyl elimination is determined both by the ancillary ligands at Ln and by the iden-tity of the alkyl ligand, and both reactions can occur for a single compound (Figure 5.21).

5.10 **Catalytic reactions**

The elementary reactions outlined in Section 5.9 can be combined to form catalytic cycles including both Ln and An. The lability of f-element to ligand bonds means that very high catalytic turnovers (much higher than achieved with d-transition metal catalysts) can often be achieved. Catalysis by organo-Ln complexes has been investigated very thoroughly, and impressive activity and selectivity (including enantioselectivity) has been achieved.

Both activity and selectivity of a lantha-noid catalyst vary with metal ion radius, and performance in a particular reaction can be optimized by selecting the correct size of lanthanoid ion

However, despite all these impressive characteristics, and many elegant and efficient reactions, the air- and moisture-sensitivity of organo-Ln complexes mean that organo-lanthanoid catalysis has not yet made the transition from lab bench to manufacturing plant.

5.10.1 Alkene hydrogenation

Alkyl-Ln species are pre-catalysts for homogeneous alkene hydrogenation: the active Ln hydride is generated *in situ* by σ-bond metathesis as shown in Figure 5.22. Where the substrate has more than one double bond, Ln catalysts are generally selective for hydrogenation of the least sterically hindered alkene. Impressive catalytic turnovers can be achieved, e.g. $[Cp^*_2LuH]_2$ gives up to 120 000 turnovers h^{-1} at 25 °C and 1 atm H_2 compared with 3000 h^{-1} achieved with $[RhCl(PPh_3)_3]$ under similar conditions.

5.10.2 Hydrosilylation

The active catalyst is an Ln hydride that is generated *in situ* from an Ln-alkyl pre-catalyst by σ-bond metathesis. The mechanism of the organo-Ln catalysed hydrosilylation reaction is analogous to that of the alkene hydrogenation reaction. The alkene insertion step is rapid, exothermic, and irreversible. This is followed by the σ-bond metathesis, which is the rate-determining step.

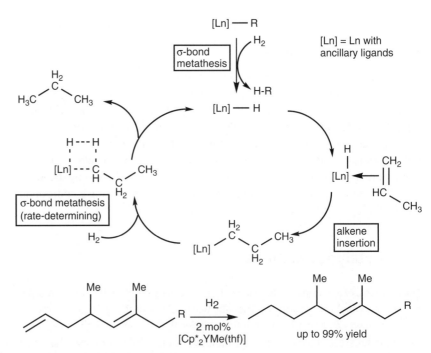

Figure 5.22 Catalytic cycle for alkene hydrogenation

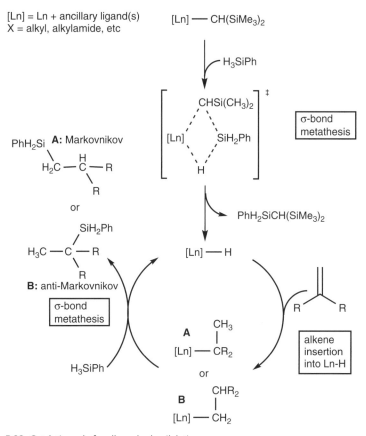

Figure 5.23 Catalytic cycle for alkene hydrosilylation

The outcome of the reaction (Markovnikov or anti-Markovnikov) is determined by the regiochemistry of the alkene insertion. High selectivity for anti-Markovnikov addition can be achieved when the catalyst has very sterically demanding ancillary ligands. The catalytic cycle for hydrosilylation is shown in Figure 5.23.

5.10.3 Intramolecular hydroamination/cyclization

The hydroamination/cyclization reaction of unsaturated amines is a useful and efficient route into N-heterocycles. The pre-catalyst for these reactions is of the type [Ln]–X where X = alkyl or NR_2, and the active catalyst is generated by reaction of the pre-catalyst with the amine substrate as shown in Figure 5.24. This reaction was first investigated using Cp_2Ln–X pre-catalysts, but has now been extended successfully to non-Cp ancillary ligands. High enantioselectivity can be achieved with the correct choice of chiral ancillary ligand, e.g. the binaphthyl-derived ligand shown in Figure 5.24.

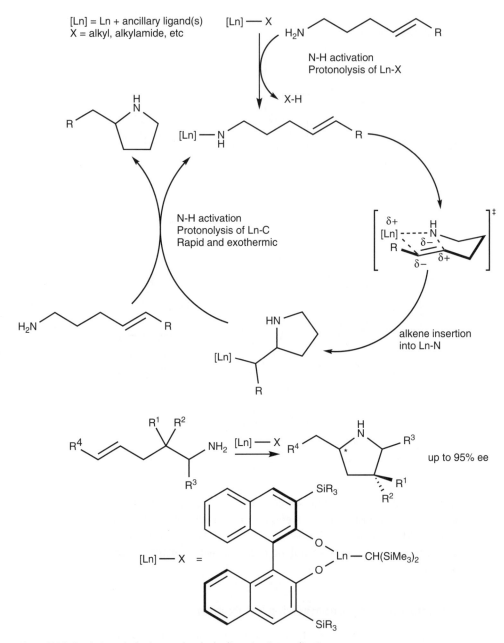

Figure 5.24 Catalytic cycle for intramolecular hydroamination cyclization

5.11 **Summary**

The organometallic chemistry of the f-elements is very different from that of the d-transition metals. The bonding (especially for Ln) is predominantly electrostatic, and complexes are frequently highly labile. There is an extensive chemistry with π-donor ligands, particularly Cp^-, which can stabilize oxidation states +2, +3, and +4. σ-bonded alkyls are highly reactive: their most important reactions are σ-bond metathesis, β-elimination, and insertion. These basic reactions can be combined to give a rich catalytic chemistry for both Ln and An.

5.12 Exercises

1. Predict the outcomes of the following reactions:

   (a) $[U(COT)_2] + PrCl_3$

   (b) $K[Pr(COT)_2] + UCl_4$

   (c) $K[Cf(COT)_2] + PuCl_4$

   (d) $CeCl_3 + 2K_2COT$

2. (a) What functions do ancillary ligands fulfil in organolanthanoid chemistry?

   (b) How would you prepare a sample of $[Y\{CH(SiMe_3)_2\}_3]$?

   (c) Suggest a synthetic route to $[Y(L)\{CH(SiMe_3)_2\}]$. Hint: $pK_a \approx 12$ for a typical 3,3'-substituted binaphthol.

   (d) Why is ligand L preferable to unsubstituted binaphthol as an ancillary ligand for organolanthanoid chemistry?

   (e) How would you expect $[Y(L)\{CH(SiMe_3)_2\}]$ to react with H_2?

$[Y(L)\{CH(SiMe_3)_2\}]$

5.13 Further reading

1. Gromada, J., J.F. Carpentier, and A. Mortreux, 'Group 3 metal catalysts for ethylene and alpha-olefin polymerization'. *Coordination Chemistry Reviews*, 2004, **248**(3–4): 397–410.

2. Andrea, T. and M.S. Eisen, 'Recent advances in organothorium and organouranium catalysis'. *Chemical Society Reviews*, 2008, **37**(3): 550–67.

3. Zimmermann, M. and R. Anwander, 'Homoleptic Rare-Earth Metal Complexes Containing Ln–C σ-Bonds'. *Chemical Reviews*, 2010, **110**(10): 6194–259.

4. Kaltsoyannis, N., 'Does Covalency Increase or Decrease across the Actinide Series? Implications for Minor Actinide Partitioning'. *Inorganic Chemistry*, 2013, **52**(7): 3407–13.

5. Neidig, M.L., D.L. Clark, and R.L. Martin, 'Covalency in f-element complexes'. *Coordination Chemistry Reviews*, 2013, **257**(2): 394–406.

6. Seaman, L.A., et al., 'In Pursuit of Homoleptic Actinide Alkyl Complexes'. *Inorganic Chemistry*, 2013, **52**(7): 3556–64.

7. Ephritikhine, M., 'Recent Advances in Organoactinide Chemistry as Exemplified by Cyclopentadienyl Compounds'. *Organometallics*, 2013, **32**(9): 2464–88.

8. Evans, W.J., 'Tutorial on the Role of Cyclopentadienyl Ligands in the Discovery of Molecular Complexes of the Rare-Earth and Actinide Metals in New Oxidation States'. *Organometallics*, 2016, **35**(18): 3088–100.

6 Applications

6.1 Introduction

The unique properties of lanthanoids and actinoids that are described in earlier chapters have hinted at many potential applications of the elements and their compounds. There have been many examples of clever and elegant chemistry (e.g. homogeneous catalysis) that really *ought* to be useful, but have not for various reasons made the transition to widespread use. This chapter will focus on actual (as opposed to potential) applications, illustrating properties that have been described earlier.

The charts in Figure 6.1 show the main applications of the lanthanoids and Y, from which it can be seen that very large quantities of La and Ce (the most abundant of the lanthanoids) are used in metallurgy and glass polishing. Other large-scale applications of the lanthanoids are in rare earth magnets and in alloys for rechargeable Ni-metal-hydride batteries. Some lanthanoids have relatively high abundance, but no large-scale applications, leading to an imbalance of supply and demand, which is a constant challenge in the economics of lanthanoid production (see Chapter 7).

The gas-mantle, containing Ce-doped ThO_2, made gas lighting viable. Invented in 1884, it was the first commercial application of f-elements.

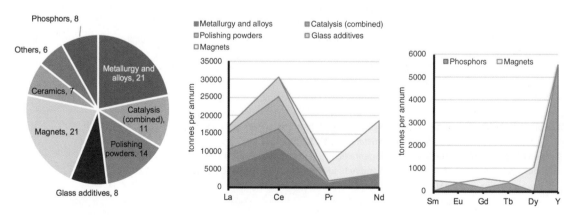

Figure 6.1 Applications of rare earths

The main commercial use of Eu and Tb is in phosphors. This is a high-value application that makes use of the unique optical properties of these elements (see Chapter 2).

The largest application of the actinoids is in nuclear fuels (in particular ^{235}U). However, there are other smaller-scale applications, for example ^{241}Am is used in smoke detectors, and ^{238}Pu is used in long-lived batteries for space probes. Depleted uranium (which is approximately 99.7% ^{238}U) is used in armour plating and armour piercing artillery due to its high density (19.5 g cm^{-3} cf 11.35 g cm^{-3} for Pb), and uranium in the form of uranium oxo-anions is used as a pigment in green 'uranium glass'.

A luminescent Eu β-diketonate complex is used in the security inks on Euro banknotes (see Chapter 4)

6.2 Lanthanoids in heterogeneous catalysis

6.2.1 Fluid catalytic cracking of petroleum (FCC)

FCC converts high boiling, high molecular weight petroleum into more useful lower boiling, lower molecular weight fractions. The main catalyst for this process is a zeolite (faujasite) in which H_3O^+ and Na^+ have been exchanged for Ln^{3+} giving a Ln^{3+} content of approximately 2.5% by mass. The exchange is achieved by soaking the zeolite in an aqueous solution of Ln^{3+}, and the result is a stabilized zeolite that is resistant to high-temperature sintering, and retains availability of catalytically active tetrahedral Al sites. Ln^{3+} may be La^{3+} or Ce^{3+}, or a mixture; La^{3+} is more effective, but the choice is often based on the relative prices of La and Ce.

6.2.2 CeO$_2$ in automotive catalytic converters

The catalyst used in automotive catalytic converters consists of CeO_2 (or a solid solution of Ce/ZrO_2) and platinum group metal nanoparticles (Pd, Pt, and Rh). It has three functions (hence the term 'three way catalyst' or TWC), which are:

- Oxidation of CO to CO_2
- Oxidation of hydrocarbons to $CO_2 + H_2O$
- Reduction of NO_x to $N_2 + O_2$

The platinum group metal nanoparticles catalyse all of these three reactions, and the role of CeO_2 (ceria) is as an O_2 buffer for reversible O_2 storage. CeO_2 is uniquely suited to this application due to the availability of both Ce^{4+} and Ce^{3+} oxidation states. When Ce^{4+} is reduced to Ce^{3+}, O_2 is released and O vacancies are created in the lattice (see Chapter 3).

$$2Ce_{Ce} + O_o \rightarrow V_o^{\cdot\cdot} + 2Ce'_{Ce} + \tfrac{1}{2}O_2$$

In addition to its applications in three way catalysts, ceria is one of the most effective catalysts for combustion of carcinogenic soot particulates that are produced in diesel engines. Ceria is used in two ways: it can be added to diesel fuel or it can be adsorbed onto the surface of a filter that removes particulates from

$$2NO + Ce_2O_3 \longrightarrow N_2O + 2CeO_2$$

$$N_2O + Ce_2O_3 \longrightarrow N_2 + 2CeO_2$$

$$CO + 2CeO_2 \longrightarrow CO_2 + Ce_2O_3$$

$$hydrocarbon + CeO_2 \longrightarrow CeO_{2-x} + CO_2 + H_2O$$

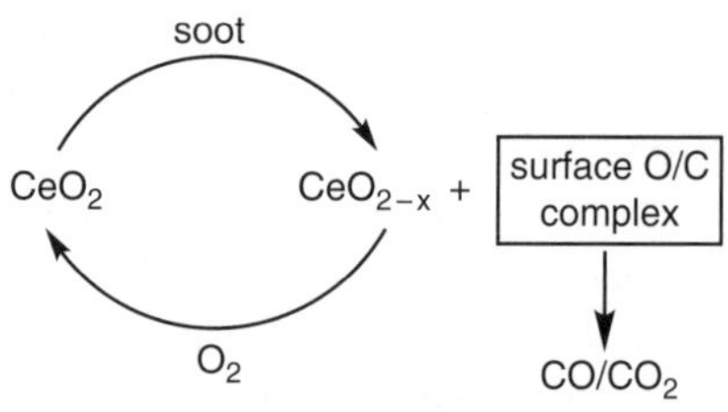

Figure 6.2 CeO_2 in automotive catalysts

the exhaust gases. Both methods work, but the latter method is less wasteful and is therefore preferred. One mechanism for soot combustion is the 'active oxygen mechanism' in which the soot particle adheres to the CeO_2 surface and surface oxygen carbon complexes (SOCs) are formed with concomitant reduction of CeO_2 to CeO_{2-x}. The SOCs produce CO_2 and CO while the CeO_{2-x} is re-oxidized to CeO_2. There is also an alternative mechanism in which NO_x act as oxidizing agents to convert soot to CO and CO_2, a process that is also catalysed by CeO_2. Figure 6.2 summarizes the roles of CeO_2 in automotive catalysts.

6.3 CeO_2 in glass polishing

One of the biggest applications of CeO_2 is in glass polishing powders that are required for production of products such as flat screen displays, high precision optics, and camera lenses. It is generally used as a slurry of sub-micron sized particles in water. In addition to having a suitable hardness compared with the glass surface, there is also a chemical contribution to the glass polishing process. Glass surfaces carry OH groups, and on exposure to CeO_2/H_2O, some O^{2-}/OH^- exchange occurs, which is accompanied by reduction of Ce^{4+} to Ce^{3+}, and the CeO_2 powder particle becomes chemically bound to the glass surface through strong Si–O–Ce bonds as shown in Figure 6.3. When the CeO_2 particle detaches from the glass, it takes a monolayer of SiO_2 with it, so there is chemical erosion of the glass surface. Rough surface features are most susceptible to this erosion and so the polishing process produces a smooth glass surface.

La$_2$O$_3$ is also used in glass polishing, but it is slightly less effective than CeO_2 due to the lack of redox chemistry.

6.4 Rare earth magnets

The rare earth magnet in a typical 3.6 MW wind turbine contains approximately 650 kg of Nd. On a smaller scale, rare earth magnets can be used in dentistry for holding dentures in place.

Rare earth magnets account for approximately 20% of total rare earth consumption and so, although not strictly chemistry, a brief account is given here.

Most of the lanthanoid metals are paramagnetic due to unpaired 4f electrons, but few of them display ferromagnetism, and only Gd is ferromagnetic at room temperature ($T_C = 293K$). This lack of ferromagnetism is due in part to the long

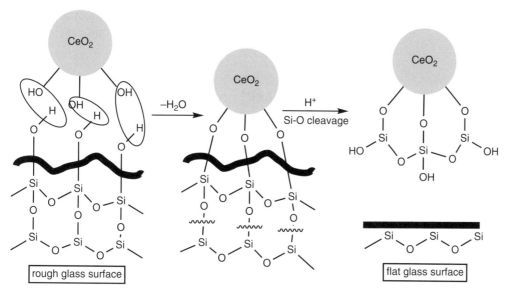

Figure 6.3 CeO_2 in glass-polishing

interatomic distances, which weaken the cooperative interactions that are necessary for ferromagnetism.

Apart from La ($4f^0$), Gd ($4f^7$), and Lu ($4f^{14}$), all of the lanthanoids have an anisotropic f-electron distribution: oblate for Nd ($4f^3$) and prolate for Sm ($4f^6$), and this, combined with strong spin-orbit coupling, is fundamental to their magnetic properties. The anisotropic electron distribution interacts with an axial crystal field so that the magnetization aligns either along ('easy axis' magnetization) or in the plane perpendicular to ('easy plane' magnetization) the unique axis. In combination with the ferromagnetic d-transition metals Fe ($3d^6$) or Co ($3d^7$) in an axial crystal field, extremely powerful permanent magnets can be produced. The three classes of rare earth magnet ($SmCo_5$, Sm_2Co_{17}, and $Nd_2Fe_{14}B$) all crystallize with the required axial symmetry.

Typical properties of the three classes of rare earth magnet are summarized in Table 6.1. The energy product BH_{max} is a measure of the maximum density of magnetic energy stored in a magnet. (For comparison, a typical AlNiCo magnet has $BH_{max} = 50$ kJ m^{-3}.) The properties of rare earth magnets can be optimized by

Table 6.1 Properties of rare earth magnets

Material	Structure	Remanence/T	Coercivity/T	Energy product BH_{max}/kJ m^{-3}	T_c/°C
$SmCo_5$	hexagonal	1	0.83	160	720
Sm_2Co_{17}	hexagonal	1.15	0.6	180	800
$Nd_2Fe_{14}B$	tetragonal	1.2	1.2	300	310

substitution of some of the d-transition metals (particularly to reduce the need for Co) and by introducing other lanthanoid metals (e.g. substituting some Nd with Dy and/or Tb).

6.5 Gadolinium in MRI contrast agents

Annual world consumption of Gd for MRI contrast agents is estimated at approximately 200 tonnes.

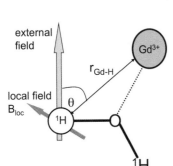

Magnetic resonance imaging (MRI) is a powerful and widely used non-invasive technique in diagnostic medicine. The image obtained in a MRI scan is based on water protons in the body (the human body consists of approximately 60% H_2O, of which approximately 43% is extracellular), making MRI particularly powerful for soft-tissue imaging. The contrast in the image depends on differences in proton relaxation times in different tissues: in the most common MRI technique protons with a short longitudinal relaxation time (T_1) show up as bright and those with a long T_1 as dark in the image.

Sometimes the natural contrast in the image is not sufficient to differentiate the tissues of interest, and in these circumstances (about 35% of MRI scans) a contrast agent is needed. The purpose of the contrast agent is to localize in specific tissues and reduce T_1 for water protons in those tissues. Gd^{3+} ($4f^7$) has the required magnetic properties: it has slow electron spin relaxation, and the dipolar interaction between the magnetic moment of Gd^{3+} and proton spins of coordinated H_2O results in shortening of T_1 for H_2O. A ^1H nucleus in a coordinated H_2O experiences both the external magnetic field of the MRI scanner and a local magnetic field B_{loc} due to dipolar interaction with the electron spin magnetic moment on Gd^{3+}. Random molecular tumbling in solution causes B_{loc} to oscillate, and when this oscillation matches the ^1H Larmor frequency, ^1H spin relaxation occurs.

However, free Gd^{3+} is toxic: it precipitates out of solution at pH 7.4 (physiological pH) and it can disrupt Ca^{2+} signalling mechanisms, and so the Gd^{3+} must be administered to the patient in the form of a stable and non-toxic complex. As we saw in Chapter 4, the most stable complexes of Gd^{3+} in aqueous solution are formed with aminopolycarboxylate ligands, and these complexes were recognized in the 1980s as potential MRI contrast agents. Figure 6.4 shows ligands that are used with Gd^{3+} in commercial MRI contrast agents and Figure 6.5 shows the use of a contrast agent in the visualization of disease. Complexes of Gd^{3+} with all of these ligands contain one inner sphere H_2O, which exchanges rapidly with bulk H_2O in solution (lifetimes are of the order of 100 ns). This rapid exchange is essential in order to minimize the dose of contrast agent required; the usual level is 0.1 mmol per kg of body weight. The contrast agent needs to be excreted from the body as soon as possible after the MRI scan is complete (see Table 6.2), and this occurs via normal renal excretion (leading to some concerns about Gd contamination in waste water). Gd^{3+} complexes with $DTPA^{5-}$ and $DOTA^{4-}$ are negatively charged, and in order to minimize disruption to osmotic pressure in the body, it is preferable to use neutral complexes (one particle per mol compared with at least two for a salt), and this is achieved in Omniscan and ProHance.

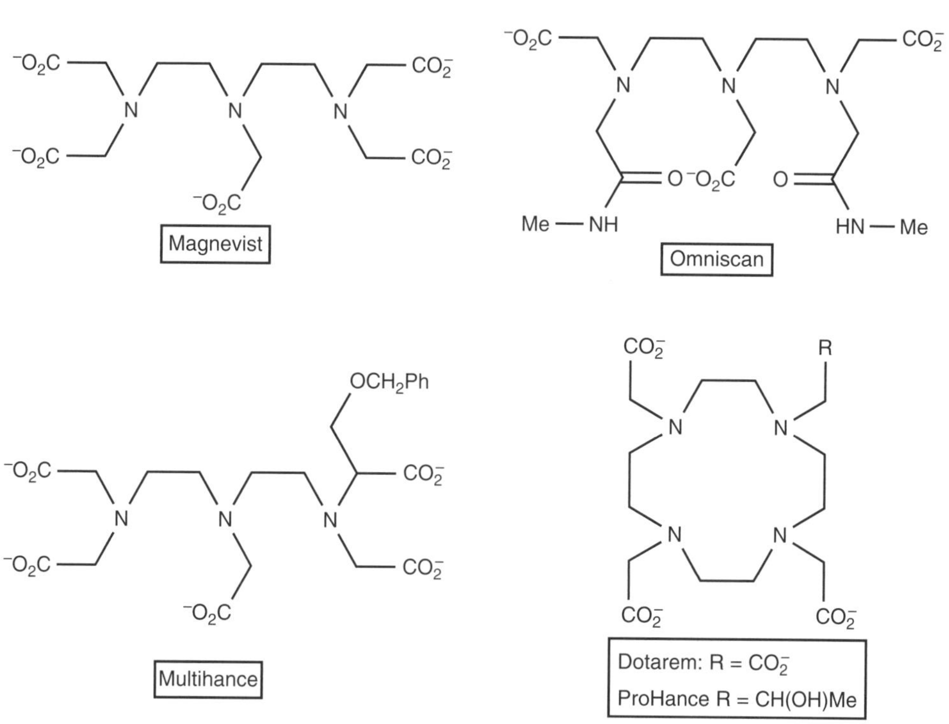

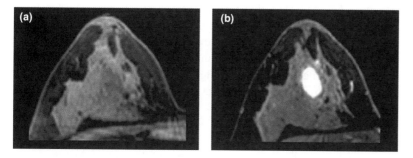

Figure 6.4 Ligands used with Gd^{3+} in MRI contrast agents

(a)

(b)

Figure 6.5 MRI contrast agents in the visualization of disease: (a) MRI of breast without contrast agent; (b) MRI of breast after injection of 0.1 mmol kg^{-1} of $[Gd(DTPA)]^{2-}$ showing fibroadenoma (bright area). (R. Weisskoff, Epix Medical, Cambridge MA)

The effectiveness of MRI contrast agents is quantified as relaxivity (see Table 6.2), which varies with magnetic field:

$$\text{relaxivity} = \frac{\Delta\left(\dfrac{1}{T_1}\right)}{\text{concentration of contrast agent}} \quad \text{where } \Delta\left(\frac{1}{T_1}\right) \text{ is change in } T_1$$

Table 6.2 Properties of MRI contrast agents

Contrast agent	T_1 relaxivity in plasma at 1.5 T (mM^{-1} s^{-1})	Half-life in body (h)
Magnevist	4.1	1.6
Omniscan	4.3	1.3
Multihance	6.3	1.17
Dotarem	3.6	1.5
ProHance	4.1	1.57

6.6 Rare earth phosphors in lighting and displays

Approximately 8000 tonnes per annum of rare earths is used in phosphors for lighting and displays. The best colour rendering is obtained using a combination of red, green, and blue emitters that are matched closely to the sensitivity of the cone cells in the human eye, and the emissions from Eu^{3+} (red) and Tb^{3+} (green) meet the required criteria. The sharpness of f→f transitions (see Chapter 2) results in particularly brilliant emissions. For applications in phosphors, luminescent Eu^{3+} and Tb^{3+} ions are doped at low levels into a host lattice (often La_2O_3 or Y_2O_3). However, as we saw in Chapter 2, 4f→4f transitions are forbidden by the Laporte selection rule and so direct excitation to the emissive states is very inefficient and alternative methods of excitation are required.

Eu^{3+} (which is easily reduced) can be efficiently excited into an emissive state via an allowed ligand to metal charge transfer (CT) transition from O^{2-} ion in the host lattice. Alternatively, indirect excitation can be achieved via valence band to conduction band transition in the host lattice. Excitation of Tb^{3+} is normally achieved via a Ce^{3+} sensitizer, taking advantage of the allowed 4f-5d transition that occurs in the UV at approx. 200 nm (Figure 6.6). Indirect excitation of Eu^{3+} and Tb^{3+} can also be achieved via the Hg emission at 254 nm.

The emission spectrum from a typical fluorescent lamp is shown in Figure 6.7.

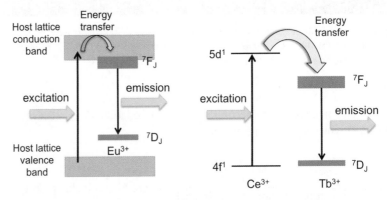

Figure 6.6 Indirect excitation of Eu^{3+} and Tb^{3+} phosphors

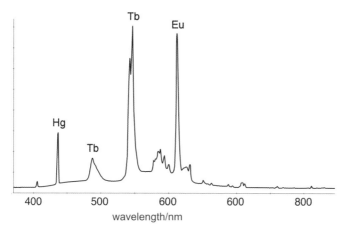

Figure 6.7 Emission from a typical fluorescent lamp

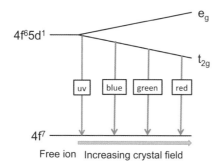

Figure 6.8 Tunable emission from Eu^{2+} phosphor

Eu^{2+} is used widely in blue phosphors, particularly in displays. Excitation of Eu^{2+} ($4f^7$) can be achieved via the Laporte allowed $4f^7 \rightarrow 4f^6 5d^1$ transition. The 5d orbitals are strongly affected by the crystal field of the matrix: the softer the anions in the matrix, the larger will be the crystal field splitting as shown in Figure 6.8. This means that the choice of host matrix determines the colour of the emission. For example when Eu^{2+} is doped into $BaMgAl_{10}O_{17}$ (BAM) the emission is in the blue; doping into SrS results in a very large crystal field splitting and a red emission. The thermal stability of BAM is poor and so although Eu^{2+}:BAM is an excellent blue phosphor it is not used in all fluorescent lamps. 5d–4f emission from Eu^{2+} is much broader than the 4f–4f emissions from Eu^{3+} and Tb^{3+}.

6.7 Luminescent lanthanoid complexes in bioassays

The complexes of Eu^{3+} and Tb^{3+} with the bipyridyl derived cryptand ligand bipy′(2.2.2) are inert in aqueous solution and strongly luminescent with relative long luminescent lifetimes (see Chapter 2 and Chapter 4). These properties make them ideal for use in time-resolved fluorescent (TRF) bioassays as shown in Figure 6.9.

$[Ln\{bipy′(2.2.2)\}]^{3+}$

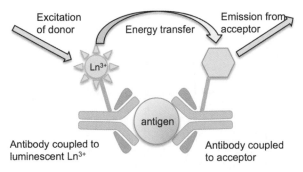

Figure 6.9 Time resolved fluorescence assay

Antibodies that are specific to the antigen under investigation are functionalized with either the luminescent Ln^{3+} cryptate or an acceptor molecule that can accept energy from Ln^{3+} fluorescence. In the presence of antigen, the donor- and acceptor-functionalized antibodies are brought into close proximity by binding to the antigen. When the sample is irradiated in the UV, Ln^{3+} is efficiently excited into its fluorescent state via the antenna effect. The close proximity of the acceptor means that energy can be transferred from Ln^{3+} to the acceptor, which then itself fluoresces. Emission from the acceptor can only be observed in the presence of the antigen: without the antigen Ln^{3+} and acceptor will not come close enough together for the necessary energy transfer to occur. The emission from the acceptor is at a lower energy than that from Ln^{3+} and so it is easy to distinguish between the two emissions. UV irradiation of sample solutions results in appreciable background fluorescence, which is usually short-lived. This can be eliminated from the experimental measurements by introducing a delay between excitation and measurement of fluorescence. The relatively long excited state lifetime of the Ln^{3+} cryptate is an advantage here.

6.8 The Nd:YAG laser

The Nd:YAG laser is the most widely used IR laser, with applications in medicine, dentistry, and manufacturing. It consists of Nd^{3+} ions doped into yttrium aluminium garnet (YAG; $Y_3Al_5O_{12}$), replacing approximately 1% of the Y^{3+} in the host lattice.

Laser emission involves a transition from a metastable excited electronic state (the emissive state) to a lower energy state (the terminal state), and requires that the population of the emissive state is greater than the population of the terminal state. If the terminal state is the ground state, then more than half of the ions must be in the emissive state, which is not a trivial situation to achieve. However, if the terminal state has higher energy than the ground state, its population will be very small and determined just by the Boltzman thermal equilibrium. Achieving a higher population of the emissive state with respect to this terminal state therefore becomes a much easier task, especially if there is efficient non-radiative decay from the terminal state to the ground state. This is

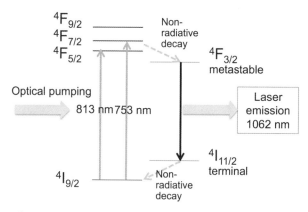

Figure 6.10 Nd^{3+} four-level laser

the principle of the four-level laser, and the Nd^{3+} ion has all the required properties. The wavelength of emission from the Nd^{3+} laser is 1.06 µm, in the infrared; a
schematic energy level diagram is given in Figure 6.10.

Optical pumping excites the Nd^{3+} ions into the ^{4}F manifold. There is then fast
non-radiative decay to the metastable $^4F_{3/2}$ state, which is the emisssive state,
giving a population inversion between $^4F_{3/2}$ and $^4I_{11/2}$ (the terminal state). The
$^4F_{3/2}$ to $^4I_{11/2}$ transition is the laser emission. The $^4I_{11/2}$ terminal level is approximately 2000 cm^{-1} above the $^4I_{9/2}$ ground state and at room temperature its fractional population is approximately 10^{-4}.

6.9 Rare earths in alloys for rechargeable Ni/MH batteries

Nickel/metal-hydride rechargeable batteries are safe and relatively cheap
compared with Li–ion batteries, and they are widely used in electric and hybrid
vehicles. The principles of the Ni/MH battery are summarized in Figure 6.11.

The most widely adopted material for the negative electrode is an alloy of
composition AB$_5$ in which A is a lanthanoid and B is a d-transition metal (mainly
Ni). A key requirement of the alloy is that it can reversibly take up and release

$$Ni(OH)_2 + OH^- \underset{\text{discharge}}{\overset{\text{charge}}{\rightleftharpoons}} NiOOH + H_2O + e^- \quad \text{positive electrode}$$

$$M + H_2O + e^- \underset{\text{discharge}}{\overset{\text{charge}}{\rightleftharpoons}} MH + OH^- \quad \text{negative electrode}$$

$$Ni(OH)_2 + M \underset{\text{discharge}}{\overset{\text{charge}}{\rightleftharpoons}} NiOOH + MH \quad \text{overall}$$

Figure 6.11 Redox reactions in the Ni/MH battery

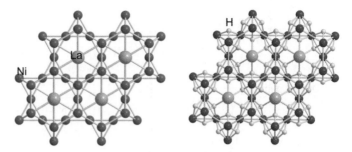

Figure 6.12 Structures of LaNi$_5$ (left) and LaNi$_5$H$_6$ (right)

hydrogen. LaNi$_5$ is able to take up 6 hydrogen atoms to form LaNi$_5$H$_6$, resulting in a higher H-atom density than that of liquid H$_2$ and a theoretical electrochemical capacity of 372 mAh/g. Figure 6.12 shows the structures of LaNi$_5$ and LaNi$_5$H$_6$.

LaNi$_5$ degrades on continued charge/discharge cycles and is therefore not suitable for use in practical applications. Its performance can be improved by replacing some of the Ni, mainly with Co, and replacing some of the La with other lanthanoids. A typical anode material now has a composition LnNi$_{3.55}$Co$_{0.75}$Mn$_{0.4}$Al$_{0.3}$ (Ln represents mischmetal with the composition La$_{0.62}$Ce$_{0.27}$Pr$_{0.03}$Nd$_{0.08}$). The battery in a Toyota Prius hybrid car contains approximately 10–15 kg of lanthanoid metal, and battery alloys is now one of the major applications by quantity of lanthanoids.

6.10 Nd in polymerization catalysis: synthesis of 1,4-cis polybutadiene (Nd-BR)

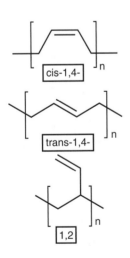

Polybutadiene is the world's second most important synthetic rubber by volume of production. There are three possible isomers: cis-1,4-, trans-1,4-, and 1,2-, of which cis-1,4- is the most useful due to its high elasticity. The major application ($\approx$ 70%) of cis-1,4-polybutadiene is in high performance 'green' car tyres that have optimized rolling resistance (thus increasing fuel efficiency) and good resistance to abrasion. A much smaller volume application ($\approx$ 1%) is in golf ball cores, which account for 30 000 tonnes per annum.

The best quality cis-1,4-polybutadiene (up to 99% cis-1,4-, highly linear, and useful molecular weight distribution) is produced using a Nd-based catalyst system (Figure 6.13), and is known as Nd-BR. The catalyst is generated *in situ* from an Nd carboxylate with a halide donor and an Al alkyl. Catalytic activity has been examined for the whole Ln series, and shows a marked dependence on Ln^{3+} radius: Nd is the most effective of the early Ln, and there is a rapid decline in activity from Gd to Lu. Sm and Eu have virtually no catalytic activity: this is because they are reduced to the +2 oxidation state by Al^{i}Bu$_2$H. The large ionic radius of Nd^{3+} allows the butadiene to chelate to the catalytic centre, and this is crucial for the exclusive formation of cis-1,4-polybutadiene.

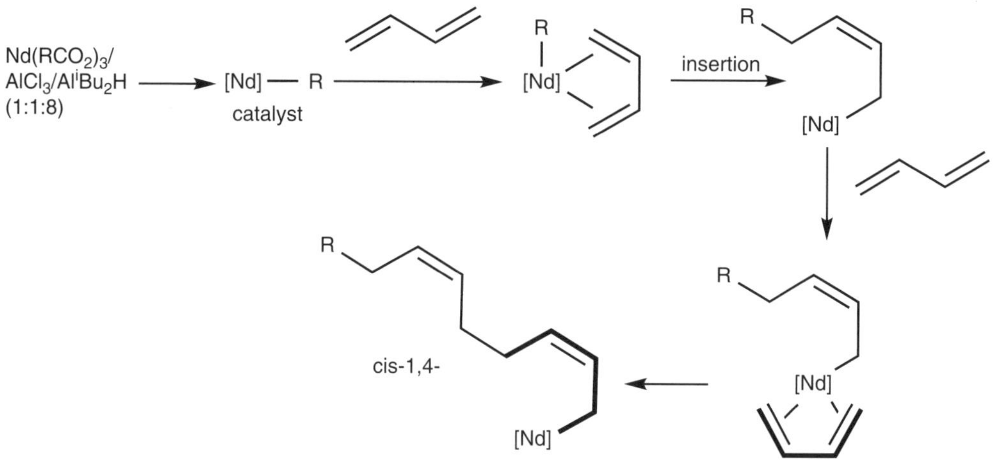

Figure 6.13 Polymerization of butadiene to form cis-1,4-polybutadiene

6.11 Lanthanoids as catalysts and reagents for organic synthesis

As described in Chapter 5 many organolanthanoid complexes are highly active as homogeneous catalysts for organic transformations, but, apart from Nd catalysis of butadiene polymerization, homogeneous Ln catalysts have not been adopted for industrial processes. However, many Ln catalysts and reagents are widely used on a laboratory scale, and a selection of the most important is given here.

6.11.1 Lewis acid catalysis

Lanthanoid triflates ($Ln(OTf)_3$; $OTf^- = CF_3SO_3^-$) were first recognized in the early 1990s as alternatives to traditional Lewis acid catalysts such as BCl_3, $TiCl_4$, $SiCl_4$, that are used widely in organic synthesis. Compared with more traditional Lewis acids, $Ln(OTf)_3$ have the advantage of forming very labile Ln–ligand bonds: this means that dissociation of product is facile, resulting in high catalytic turnovers and correspondingly low catalyst loadings (<5 mol%). The traditional Lewis acid catalysts form strong bonds to both substrate and product and so are usually required in stoichiometric quantities. Another advantage of $Ln(OTf)_3$ is that they are water-tolerant, whereas the traditional Lewis acids are rapidly hydrolysed in the presence of even small quantities of water. $Yb(OTf)_3$ is the most Lewis acidic of the lanthanoid triflates: the $4f^{13}$ configuration of Yb^{3+} makes it more Lewis acidic than Lu^{3+} ($4f^{14}$) despite having a slightly larger ionic radius. Some examples of $Ln(OTf)_3$-catalysed reactions are shown in Figure 6.14.

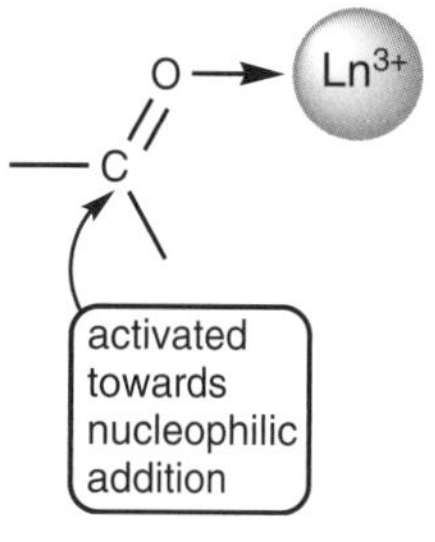

activated towards nucleophilic addition

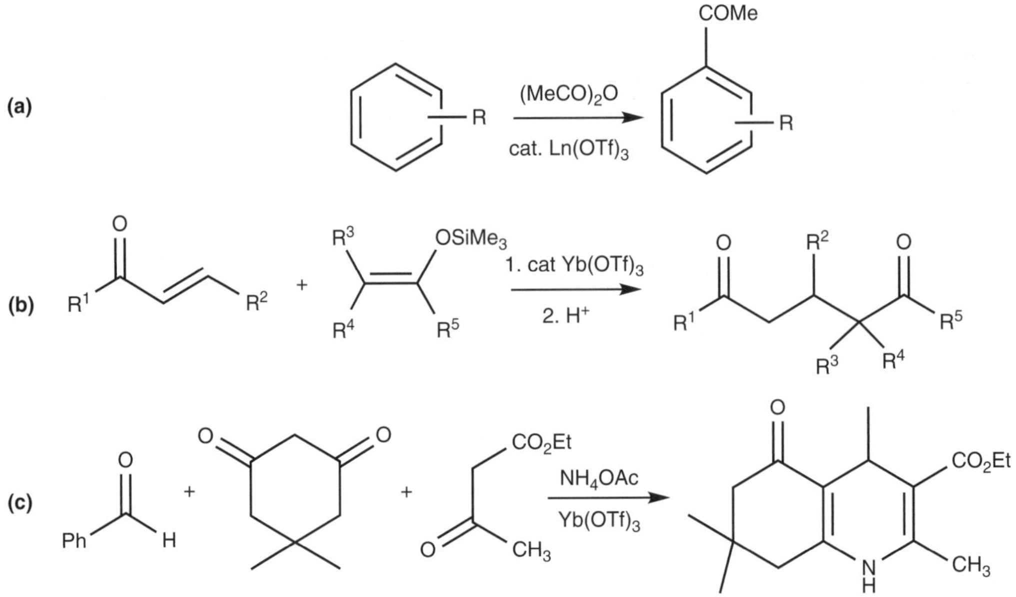

(a)

(b)

(c)

Figure 6.14 Reactions catalysed by Ln(OTf)$_3$ (a) Friedel–Crafts acylation; (b) Michael addition; (c) Hantzsch condensation

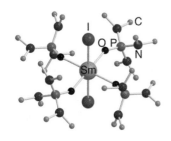

[SmI$_2$(hmpa)$_4$]

There are a few cases where SmI$_2$ can be used in catalytic quantities in combination with a co-reductant (e.g. Mg or electrochemical).

6.11.2 SmI$_2$ as a one-electron reducing agent

SmI$_2$ is soluble in tetrahydrofuran (thf) (up to 0.1 M) and can be prepared readily by reaction of metallic Sm with diiodoethane as shown in Figure 6.15.

The solubility in thf makes SmI$_2$ a useful homogeneous one-electron reducing agent, and the oxophilicity of both Sm^{2+} and Sm^{3+} means that it can be more selective (particularly with carbonyl substrates) than heterogeneous alternatives. The reducing power can be increased by addition of strong donor ligands such as (NMe$_2$)$_3$P=O. E°(Sm^{3+}/Sm^{2+}) = −1.33 V for [SmI$_2$(thf)$_n$] and −2.05 V for [SmI$_2${O=P(NMe$_2$)$_3$}$_4$].

SmI$_2$ is generally required in stoichiometric quantities, and so its use is currently limited to laboratory scale reactions. An example is the pinacol coupling reaction shown in Figure 6.15, in which high levels of diastereoselectivity can be achieved.

$$Sm + ICH_2CH_2I \xrightarrow[\text{inert atmosphere}]{\text{thf}} [SmI_2(thf)_n] + H_2C=CH_2$$

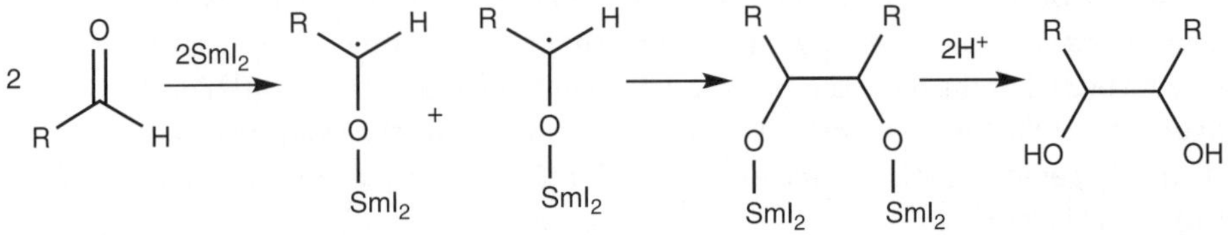

Figure 6.15 Pinacol coupling reaction promoted by SmI$_2$

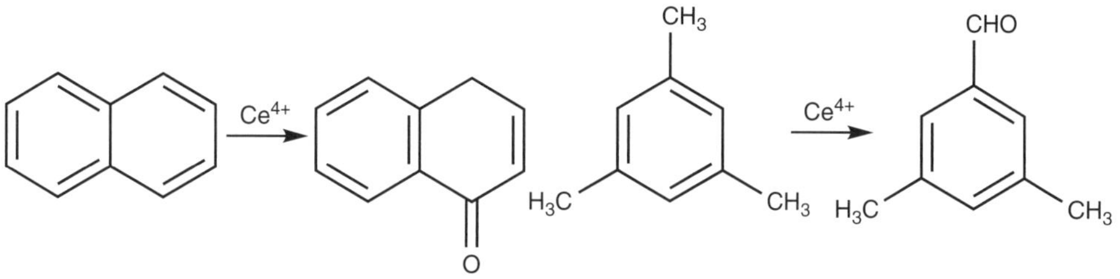

Figure 6.16 Oxidation reactions promoted by Ce^{4+}

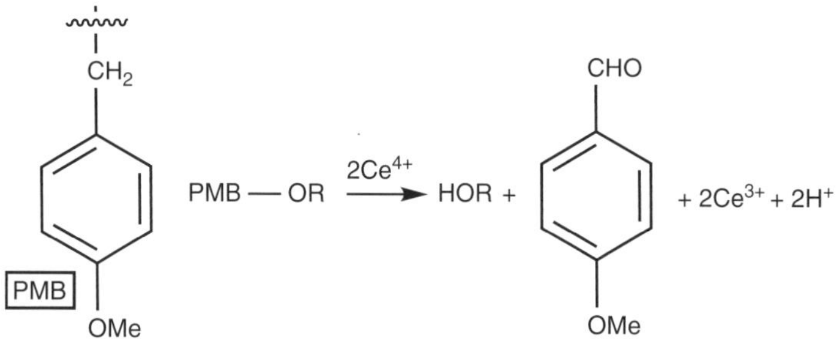

Figure 6.17 De-protection reaction promoted by Ce^{4+}

6.11.3 Ce(IV) as a one-electron oxidizing agent

Ce^{4+} is a powerful one-electron oxidizing agent (E° Ce^{4+}/Ce^{3+} = 1.72 V) and is readily available in the form of ceric ammonium nitrate [NH$_4$]$_2$[Ce(NO$_3$)$_6$] (CAN) or ceric ammonium sulfate [NH$_4$]$_4$[Ce(SO$_4$)$_4$] (CAS). CAN and CAS are both kinetically stable in acidic aqueous solution and are normally used in mixed organic/aqueous solvents such as MeCN/H$_2$O. Two examples of Ce^{4+} oxidations are shown in Figure 6.16.

Ce^{4+} is also used to remove the PMB protecting group from an alcohol as shown in Figure 6.17.

6.11.4 Organocerium reagents

Organocerium reagents were first reported in the 1980s, and they have become established alternatives to organo-Li and Grignard reagents for addition reactions to carbonyls and imines, particularly where the substrate is either sterically hindered or susceptible to enolization.

Organocerium reagents are generated *in situ*, usually from anhydrous CeCl$_3$ and the corresponding organo-Li reagent in ether solvent at −78°C. The exact nature of the reagents is not established, but despite the stoichiometry of the reaction, they are almost certainly not 'CeCl$_2$R'. The mild Lewis acidity and oxophilicity of Ce^{3+} are crucial to the selectivity of organocerium reagents. Some examples of organocerium reactions are shown in Figure 6.18.

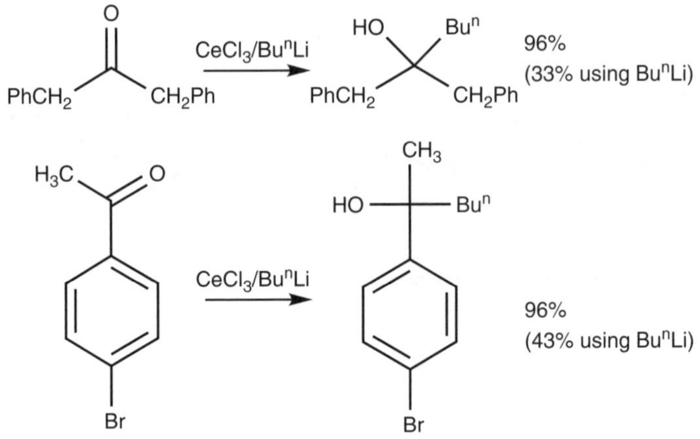

Figure 6.18 Organocerium reactions

6.12 Actinoids in nuclear power

6.12.1 Uranium

Worldwide nuclear power generation (estimated as 2 519 TWh in 2017) requires approximately 65 000 tonnes of ^{235}U per annum (equivalent to approximately 77 000 tonnes U_3O_8). ^{235}U undergoes nuclear fission as shown in Figure 6.19 when it is irradiated with 'thermal' neutrons (thermal neutrons are in thermal equilibrium with their surroundings and have energy of approximately 2 kJ mol^{-1}).

The nuclear fission reaction splits ^{235}U into two lighter fragments, initially mainly ^{141}Ba and ^{92}Kr (both of which are highly unstable and decay rapidly to other products), along with an additional one or two neutrons which can go on to promote fission of more ^{235}U nuclei in a chain reaction. The chain reaction is maintained in a 'critical' state, meaning that neutrons are produced at a constant rate. The heat generated by the fission reaction is used to convert water to steam, which then drives a turbine to generate electricity. A small proportion of the heat generated from a nuclear reactor comes from decay of radioactive fission products.

The most widely used nuclear reactor is the pressurized water reactor (PWR) (see Figure 6.20). The fuel for a PWR is UO_2 formed into cylindrical ceramic pellets (approximately 1 cm × 1.5 cm) packed into a long metal tube to form a fuel rod. The metal used for the fuel rods is an alloy of zirconium (zircaloy, which consists

1.7 billion years ago, when natural uranium contained c. 3.1% ^{235}U, geological conditions in Oklo, Gabon, were such that a natural nuclear reactor was formed. Fission reactions continued for hundreds of thousands of years until the ^{235}U content became too low.

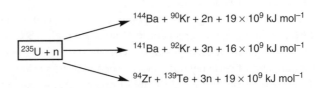

Figure 6.19 Fission reactions of ^{235}U

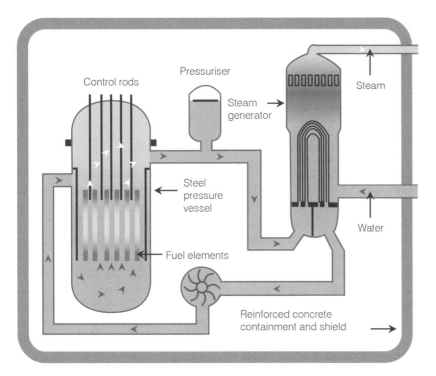

Figure 6.20 Pressurized water reactor. (World Nuclear Association; reproduced with permission)

of 98% Zr/1.5% Sn plus Fe and Cr), which is virtually transparent to neutrons. A large PWR may contain up to 70 000 fuel rods and up to 100 tonnes of uranium.

Water fulfils two functions: firstly it is the coolant that transfers heat from the core of the reactor to the steam generator, and secondly it is a moderator. A moderator is required to slow down highly energetic neutrons that are produced by the fission reaction so that they have a suitable energy (approximately 2 kJ mol⁻¹) to promote further fission reactions and sustain a chain reaction. The water in the core reaches temperatures of over 300 °C and so high pressure is required to prevent it boiling. The control rods are made of a neutron-absorbing material (e.g. boron) and they are used to control the rate of the fission reaction. The fission reaction can be stopped by lowering the control rods into the reactor core.

The fuel in a reactor degrades with time as ^{235}U is consumed and fission products build up, resulting in reduced efficiency. Approximately a quarter to a third of the fuel rods in a reactor are replaced every 12–18 months and this spent fuel is reprocessed, e.g. by the PUREX process as described in Chapter 7.

6.12.2 **Mixed oxide (MOX) fuel**

^{239}Pu is produced in nuclear reactors due to neutron capture by ^{238}U (which is the major isotope of U in nuclear fuel):

$$^{238}_{92}\text{U} + ^{1}_{0}\text{n} \rightarrow ^{239}_{92}\text{U} \xrightarrow{-\beta^-} ^{239}_{93}\text{Np} \xrightarrow{-\beta^-} ^{239}_{94}\text{Pu}$$

Like ^{235}U, ^{239}Pu is a fissile isotope, producing a similar quantity of energy to ^{235}U. Over half of the ^{239}Pu that is produced in a typical nuclear reactor is consumed *in situ*, contributing around a third of the total energy output. Spent nuclear fuel rods contain approximately 0.6% ^{239}Pu, which can be separated by the PUREX process and combined with depleted UO_2 to form mixed oxide (MOX) fuel. In 2018, MOX fuel accounted for approximately 5% of all new nuclear fuel.

MOX fuel provides a method to dispose of military (weapons-grade) plutonium

6.12.3 ^{239}Pu in fast breeder reactors

A fast neutron reactor (FNR) is a nuclear reactor that uses high energy ('fast') neutrons with energies $\geq 480 \times 10^6$ kJ mol^{-1}, and, due to the high energy of the neutrons, no moderator is required. A fast breeder reactor (FBR) is an FNR that produces more fissile material than it consumes.

A typical FBR uses MOX fuel (80% UO_2/20% PuO_2). Although fast neutrons are too energetic to sustain fission of ^{235}U, fission of ^{239}Pu can be sustained (albeit with reduced efficiency compared with thermal neutrons), and produces c. 20% more neutrons than ^{235}U fission. Fast neutrons are captured by ^{238}U to produce more ^{239}Pu. FBRs operate at 500–550 °C and atmospheric pressure, so water cannot be used as the coolant; the usual coolant is liquid sodium. Although expensive to set up, FBRs make much more efficient use of uranium resources and are able to burn actinoids that would otherwise contribute to long-lived nuclear waste.

6.13 ^{241}Am in smoke detectors

^{241}Am has a half-life of 432.6 y; it is an alpha emitter, but emits very little γ-radiation, and is used in the form of AmO_2 in ionizing smoke detectors (about 0.34 µg). Air is ionized in the ionization chamber by alpha particles, allowing an electric current to pass through as shown in Figure 6.21. When smoke particles enter the chamber, they absorb the alpha particles and the ionization process is disrupted, resulting in a drop in current, which is detected. The main source of ^{241}Am is from the reprocessing of spent nuclear fuels.

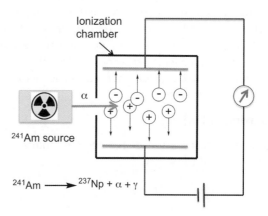

Figure 6.21 ^{241}Am smoke detector

6.14 ^{238}Pu in power supplies

^{238}Pu has a half-life of 87.4 y; it is an alpha emitter, but emits very little γ-radiation. Its relatively long half-life means that it emits alpha particles at a fairly constant rate over a significant length of time, making it suitable for use in the form of ^{238}PuO$_2$ in long-lived power sources known as radioisotope thermoelectric generators (RTGs). In the RTG, heat is generated by the alpha emission, and the heat is converted to an electric current by an array of thermocouples. RTGs are long-lived and maintenance-free and these advantages outweigh their rather low efficiency, which is generally in the range 3–7%. ^{238}Pu-containing RTGs are used in space probes such as *Cassini*, *Voyagers 1 & 2*, and *New Horizons*, in which long lifetime and reliability are essential. The *Cassini* probe had three RTGs each containing 7.8 kg of PuO$_2$, and with a maximum power output of 300 W.

Until the 1970s, RTGs were used routinely in heart pacemakers, but since then the environmental risks have been considered to be too high: ^{238}Pu can be released into the environment if bodies are cremated without prior removal of pacemakers. ^{238}Pu is not present in spent nuclear fuel and must be synthesized by neutron irradiation of ^{237}Np.

6.15 Summary

By volume, the main applications of the rare earths are in metallurgy and alloys, and in permanent magnets. The largest demand for Nd is in Nd$_2$Fe$_{14}$B permanent magnets (e.g. for use in wind turbines). The unique optical and magnetic properties of the lanthanoids are key to many of the applications of these elements. For example Eu and Tb are used as phosphors in fluorescent lighting, and the Nd:YAG laser is the most widely used IR laser. Complexes of Gd are used as contrast agents in MRI imaging.

Industrial applications of rare earths that are based on chemical properties include CeO$_2$ in automotive catalysts (the redox chemistry of Ce is crucial to this application), and Nd carboxylates as catalysts for butadiene polymerization. Rare earth catalysts and reagents are used in synthetic chemistry on a laboratory scale. There is a constant challenge in rare earth technology to find uses for the more abundant elements (La and Ce) that are inevitably produced alongside elements such as Nd that are in high demand but available in lower abundance. One major use of La$_2$O$_3$ and CeO$_2$ is in glass polishing powders.

The use of U in nuclear reactors is the main application of the actinoids. However there are smaller-scale applications based on radiochemical properties, e.g. ^{241}Am in smoke detectors and ^{238}Pu in long-lived power supplies.

6.16 Exercises

1. (a) What properties are (i) essential and (ii) desirable for a Gd^{3+} complex to be used as a MRI contrast agent?

   (b) Figure 6.22 shows a selection of water-soluble Gd^{3+} species. Use the criteria that you identified in (a) to determine the suitability of these

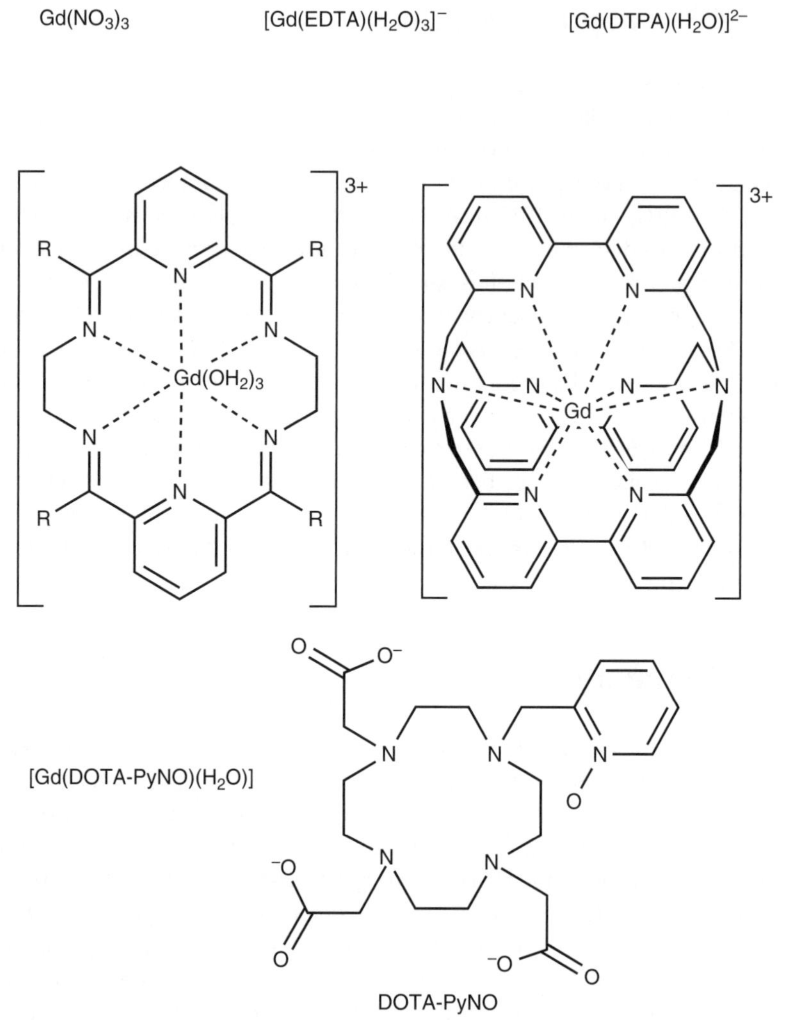

$Gd(NO_3)_3$ $[Gd(EDTA)(H_2O)_3]^-$ $[Gd(DTPA)(H_2O)]^{2-}$

$[Gd(DOTA-PyNO)(H_2O)]$

DOTA-PyNO

Figure 6.22 Structures for Exercise 1

species for use as MRI contrast agents. Are any of the species totally un-suitable? (Hint: Chapter 4 discusses coordination chemistry in aqueous solution.)

2. Why can the emission from Eu^{2+} be tuned by choice of host matrix whereas the emission from Eu^{3+} cannot?

3. $LaNi_5$ (used in nickel/metal-hydride batteries) takes up H_2 to form $LaNi_5H_6$. The unit cell of $LaNi_5H_6$ has a volume of 110.41 $Å^3$ (1 $Å = 10^{-10}$ m) and con-tains one formula unit of $LaNi_5H_6$. Calculate the density of H in $LaNi_5H_6$. (For comparison, the density of liquid H_2 at 20 K is 70.99 kg m^{-3}.)

6.17 Further reading

1. Bottrill, M., L.K. Nicholas, and N.J. Long, 'Lanthanides in magnetic resonance imaging'. *Chemical Society Reviews*, 2006, **35**(6): 557–71.

2. Moore, E.G., A.P.S. Samuel, and K.N. Raymond, 'From antenna to assay: lessons learned in lanthanide luminescence'. *Accounts of Chemical Research*, 2009, **42**(4): 542–52.

3. Procter, D.J., R.A. Flowers, and T. Skrydstrup, 2010. *Organic Synthesis Using Samarium Diiodide: A Practical Guide*, Cambridge: RSC Publishing.

4. Alonso, E., et al., 'Evaluating rare earth element availability: a case with revolutionary demand from clean technologies'. *Environmental Science & Technology*, 2012, **46**(6): 3406–14.

5. Montini, T., et al., 'Fundamentals and catalytic applications of CeO_2-based materials'. *Chemical Reviews*, 2016, **116**(10): 5987–6041.

6. Li, P., et al., 'A review on oxygen storage capacity of CeO_2-based materials: Influence factors, measurement techniques, and applications in reactions related to catalytic automotive emissions control'. *Catalysis Today*, 2019, **327**: 90–115.

7. Binnemans, K., et al., 'Rare earths and the balance problem: how to deal with changing markets?' *Journal of Sustainable Metallurgy*, 2018, **4**(1): 126–46.

Extraction and purification of lanthanoids and actinoids

7.1 Introduction

The naturally occurring actinoids U and Th were first isolated in 1789 and 1828 respectively, but isolation of individual Ln elements was a much greater challenge, finally achieved through painstaking work between 1839 (Ce) and 1945 (Pm). This chapter will cover methods for isolation of the f-elements on an industrial scale, as well as isotopic enrichment of U for use in nuclear reactors, and the reprocessing of spent nuclear fuel.

7.2 Extraction and separation of the lanthanoids and yttrium

The three most important rare earth minerals are bastnaesite ($LnF(CO_3)$), monazite ($Ln/Th(PO_4)$), and xenotime (mainly $Y(PO_4)$ in combination with $Ln(PO_4)$). All of these minerals contain mixtures of rare earth elements as shown in Figure 7.1; early Ln predominate in monazite and bastnaesite (in both of which

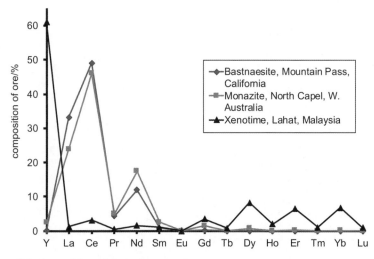

Figure 7.1 Compositions of the main rare earth ores

Ln^{3+} is 9-coordinate), whereas xenotime (in which Ln^{3+} is 8-coordinate) is richer in the late Ln elements. The presence of Th (which is radioactive) has resulted in less use of monazite in recent years due to safety considerations.

The rare earth ores all contain mixtures of the elements in the +3 oxidation state, and because the chemistry of the elements is so similar, complete separation by simple chemical means is not possible. The only property that can be exploited for separation is the difference in Ln^{3+} radius along the series, which is manifested in properties of compounds, e.g. stability constants of complexes and solubility of salts. In the early days of Ln chemistry, the elements were separated by fractional crystallization of salts, a particularly heroic example of which was the isolation in 1911 of a pure sample of Tm after a series of 15 000 crystallizations of rare earth bromates. The purity of the Tm sample was demonstrated by absorption spectroscopy, a relatively new technique at the time. Fractional crystallization is clearly not a practical method to obtain useful quantities of the pure elements, but fortunately an alternative separation method—solvent extraction—became available in the mid twentieth century, allowing the development of rare earth chemistry and technology.

Rare earth ores frequently occur alongside other minerals and so the first stage in the separation process is concentration of the rare earth-containing component. Iron-containing minerals (e.g. ilmenite, magnetite) can be removed magnetically. The next stage of the process is usually froth flotation: the mixture of minerals is first ground to a fine powder (< 1mm) and then an aqueous suspension is treated with a surfactant that binds specifically to the rare earth mineral, making the particles hydrophobic. The resulting aqueous slurry (sometimes containing a frothing agent) is aerated and the hydrophobic particles bind to the air bubbles, which rise to the surface and can be easily separated. The resulting concentrated rare earth ore then undergoes an extraction process with acid to form an aqueous solution of mixed rare earth salt. Two examples of extraction procedures are shown in Figure 7.2. Th(IV) can be separated from monazite due

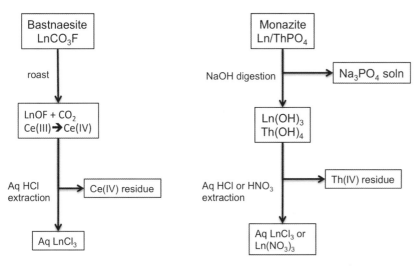

Figure 7.2 Extraction of Ln from bastnaesite and monazite

to the lower solubility in acid of $Th(OH)_4$ compared with $Ln(OH)_3$. In a similar way, Ce(IV), produced by pre-roasting of bastnaesite, can be separated from $Ln(OH)_3$.

7.2.1 Solvent extraction

The processes summarized in Figure 7.2 both result in aqueous solutions of mixed Ln^{3+} salts; the most important method for separation of these mixtures into solutions of single Ln^{3+} is solvent extraction (also known as hydrometallurgy).

In solvent extraction, an aqueous solution of metal salt is treated with an organic (usually high-boiling kerosene) solution of extractant HL. On mixing the organic and aqueous phases, a neutral, organic-soluble metal complex is formed, and this transfers into the organic phase.

The most commonly used extractant for rare earth purification is the phosphoric acid derivative di-2-ethylhexylphosphoric acid (D2EHPA). Hydrogen-bonding causes D2EHPA to form dimers in solution, and the complex formed with Ln^{3+} contains three mono-deprotonated D2EHPA dimers as ligands (Figure 7.3).

The extractant binds more strongly to the smaller, later Ln^{3+} and so, as shown in Figure 7.4(a), the later Ln^{3+} are preferentially extracted into the organic phase. The effect is small for a single stage of extraction, and so multiple extractions are required to achieve complete separation. A battery of approximately 1500 mixer-settlers is used in a typical separation plant.

Ln^{3+} is 'stripped' from the organic solution by treating with aqueous acid; the pH at which a particular Ln^{3+} is stripped depends on the Ln^{3+} radius as shown

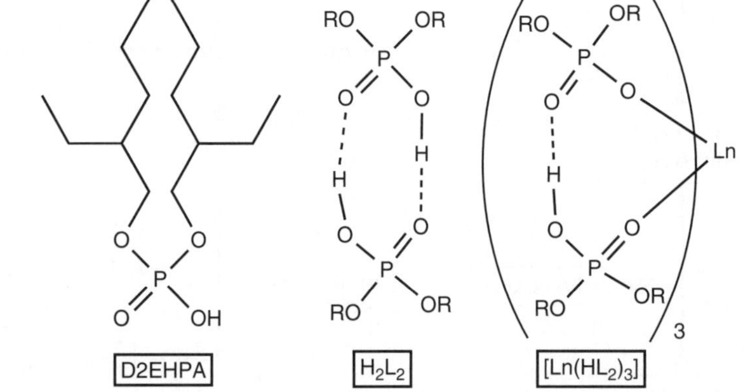

Figure 7.3 Reaction scheme for solvent extraction of Ln^{3+} using D2EHPA extractant

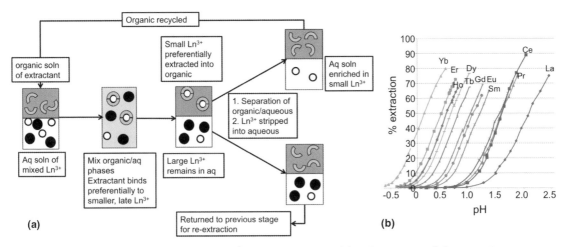

Figure 7.4 (a) Simplified scheme for separation of Ln³⁺ by solvent extraction (b) pH for stripping Ln³⁺ from organic to aqueous solution (CYANEX 572 extractant, with permission from Cytex Industries Inc.)

$$Ln^{3+}_{(aq)} \xrightarrow{H_2C_2O_4} Ln_2(C_2O_4)_2 \text{ (ppt)} \xrightarrow[\text{heat}]{\text{air}} Ln_2O_3 \xrightarrow{HF} LnF_3 \xrightarrow[1450°C]{Ca} Ln + CaF_2$$

Figure 7.5 Production of Ln_2O_3 and metallic Ln

in Figure 7.4(b). For example at pH 2.5, 80% of La³⁺ is stripped into the aqueous solution, but none of the other Ln³⁺; at pH 0, 40% of Yb³⁺ is stripped into the aqueous solution along with ca. 10% of Er³⁺ but no other Ln³⁺.

The aqueous solution of purified Ln³⁺ is treated as shown in Figure 7.5 to obtain either Ln oxide (the main form in which Ln are supplied) or metallic Ln.

7.2.2 Ion exchange chromatography

Ion exchange chromatography is significant historically as the first practicable method for Ln³⁺ separation, and it is still important for separation of the late An³⁺. However, it is not ideally suited to large-scale separations so is not important on an industrial scale. The method is shown schematically in Figure 7.6.

An aqueous solution of the mixed metal salt is loaded onto the ion exchange column, which is packed with a resin such as Dowex 50. The M³⁺ ions bind to the resin with different affinities, depending on ionic radius: larger ions have a greater affinity for the resin than small ions. The column is then eluted with a complexing agent such as $[NH_4]_3[HEDTA]$ or $[NH_4][\alpha\text{-HIBA}]$ (α-HIBA = α-hydroxyisobutyrate) and the $[NH_4]^+$ displaces the metal ions, which then bind to the complexing agent. The stability constant of the metal complex varies with M³⁺ radius so that smaller M³⁺ form more stable complexes, which are eluted from the column first. The combined effect of decreasing affinity for the resin and increasing stability constant for complex as M³⁺ radius decreases means that good separation can be achieved as shown in Figure 7.6.

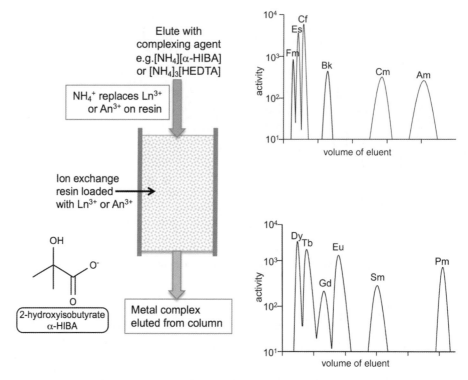

Figure 7.6 Ion exchange separation of Ln^{3+} and An^{3+} mixtures

7.2.3 Economics of rare earth element extraction

The separation of individual rare earth elements from natural ores is just one consideration in the economics of rare earth production: another major challenge is that any single ore will provide a mixture of at least seven rare earth elements, and the relative abundance of the individual elements is unlikely to mirror precisely the relative demand for these elements. For example, Nd is in high demand due to its application in magnets, but for every tonne of Nd that is extracted from bastnaesite (rich in early lanthanoids), 7.3 tonnes of La, Ce, Pr, and Sm are also extracted, but there is only demand for approximately 3.8 tonnes of this.

7.2.4 Recycling of rare earth elements

Demand for rare earth elements has increased dramatically during the twenty-first century (see Chapter 6 for applications). This has been driven by the need for permanent magnets (containing Nd and Dy) in wind turbines, alloys (containing La, Ce, Nd, and Pr) for rechargeable Ni-metal hydride batteries in hybrid cars, and an increasing use of Eu, Tb, and Y in energy-efficient fluorescent lamp phosphors. This increase in demand coincided with China (the world's major rare earth producer) imposing strict export quotas in 2012, resulting in price increases and concerns about future availability of the elements. The European Commission in 2010 and the U.S. Department of Energy in 2011 identified Nd, Eu, Tb, Dy, and Y as critical elements with high importance for clean energy, combined with high risk to future supply. Recycling of phosphors and rare earth magnets (the major applications of these critical elements) is being actively investigated.

It is possible to physically separate the phosphor (Eu^{3+} and Tb^{3+} doped into Y_2O_3) from a fluorescent lamp, but there will usually be some contamination with glass and Hg, so the recovered phosphor will inevitably be of lower quality than the original material.

There are a few cases (for large and easily accessible magnets) where it is possible to recycle an entire rare earth magnet. However, it is much more common that the magnetic alloy has to be separated from other components and then reprocessed into a new magnet. The separation is achieved by treatment with hydrogen, which breaks down the magnet into fragments that can be separated from other components and then reprocessed either by sintering or by binding in a resin matrix.

The physical separation of phosphors or magnetic alloys inevitably results in reduced quality in the recycled product; if high quality is required, then a hydro-metallurgical process (similar to extraction of the elements from the ore) must be applied. In the current economic climate (2018) recycling of rare earths is not an economically viable process.

7.3 Extraction of the actinoids

Only two actinoid elements (Th and U) occur naturally to any great extent. The main source of Th is monazite, in which it occurs alongside Ln (see Section 7.2). Pa occurs naturally in extremely small quantities from decay of ^{235}U, and the largest quantity of Pa ever isolated (130 g) was obtained in the 1960s by repeated extraction of 60 tonnes of pitchblende residue. Some trans-uranium elements (particularly Np, Pu, Am, and Cm) are formed by neutron capture in nuclear reactors and are isolated by reprocessing of spent nuclear fuels (see Section 7.5).

7.3.1 Extraction of uranium

Figure 7.7 shows the distribution of uranium ores and annual production of uranium by country. The main ore of uranium is pitchblende (also known as uraninite), which contains a mixture of UO_2 and U_3O_8 in which uranium exists in the +4 and +6 oxidation states. Pitchblende derives its name from the German 'pech', meaning 'bad luck': miners who found this U-containing mineral in the silver-rich Schneeberg region of Germany were considered to be unfortunate. Pitchblende usually co-exists with up to ten other associated minerals, so some form of pre-concentration of the ore is usually applied (e.g. magnetic separation to remove iron-containing minerals). The next stage of the procedure is to extract uranium into aqueous solution, and this is normally achieved by leaching

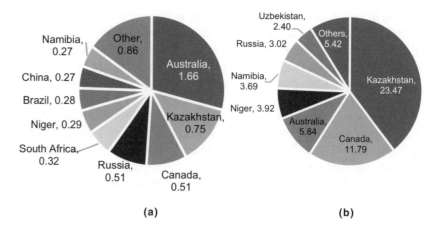

Figure 7.7 (a) U resources (Mt) and (b) Annual U production (kt)

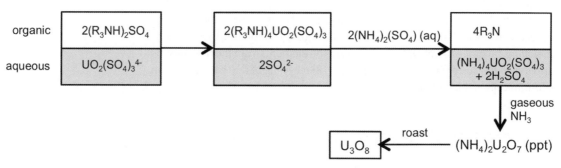

Figure 7.8 Extraction of U_3O_8 from pitchblende

with sulfuric acid. About half of all uranium is extracted by *in situ* leaching, which is associated with potential environmental risks. The aqueous solution is then oxidized to ensure that all of the uranium is in the form of UO_2^{2+} and the uranium is separated out either by solvent extraction (as shown in Figure 7.8) or by ion exchange. The final product (usually referred to as 'yellow cake') is an impure form of U_3O_8.

A further solvent extraction procedure using tri-n-butyl phosphate (TBP; $(Bu^nO)_3P{=}O$) is required to obtain the high purity of uranium that is needed for nuclear applications.

7.4 Isotopic enrichment of uranium

The major application of uranium is in nuclear reactors, and this requires isotopic enrichment to ca. 3–5% ^{235}U from natural abundance (ca. 0.7% ^{235}U). All of the technologies for isotopic enrichment use UF_6 in the gas phase: its key advantages

$$UO_2 + 4HF \longrightarrow UF_4 + 2H_2O \xrightarrow{F_2} UF_6$$

Figure 7.9 Synthesis of UF_6

are firstly that it has reasonable volatility (it sublimes at 56.5 °C at atmospheric pressure) and secondly that F is monoisotopic, so that isotopologues of UF_6 differ only in the isotope of U. The most widely used preparative route to UF_6 is shown in Figure 7.9.

7.4.1 Gas diffusion

Gas diffusion was the original technology for isotopic enrichment but due to high energy and capital costs, this method is no longer used. The method is based on the differential rates of diffusion of gaseous $^{235}UF_6$ and $^{238}UF_6$ through a semipermeable membrane, which is usually Ni or Al with pores of 10–25 nm. Graham's Law states that the rate of diffusion is proportional to $1/\sqrt{M_r}$ so that:

$$\frac{\text{rate}(^{235}UF_6)}{\text{rate}(^{238}UF_6)} = \frac{\sqrt{M_r(^{238}UF_6)}}{\sqrt{M_r(^{235}UF_6)}} = 1.0043$$

Only a very small degree of enrichment can be achieved with a single stage of gas diffusion. In order to enrich from natural abundance UF_6 (0.7% ^{235}U) to fuel grade (3.5% ^{235}U), 375 stages are required, and to obtain weapons grade (90% ^{235}U) requires 1100 stages.

$$\frac{3.5}{0.7} = 1.0043^n$$

$$\log\frac{3.5}{0.7} = n \times \log 1.0043$$

$$n = \frac{0.69897}{0.0018635} = 375$$

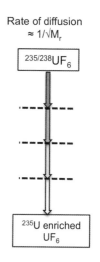

7.4.2 Gas centrifuge

Gas centrifugation is now used for virtually all uranium enrichment. It was first developed during the 1940s and came into commercial use in the 1960s. This technology is relatively simple: it consists of an evacuated casing (ca. 2m × 0.1m) containing a cylindrical rotor. The pressure within the casing is ca. 0.001 atm in order to keep the UF_6 in the gas phase, and the rotor speed is ca. 7000 rpm. Centrifugal force pushes the heavier $^{238}UF_6$ towards the wall, concentrating lighter $^{235}UF_6$ in the centre. The enriched and depleted gas streams can be separated, and a typical centrifuge achieves a separation factor of ca. 1.2, much larger than achieved using gaseous diffusion.

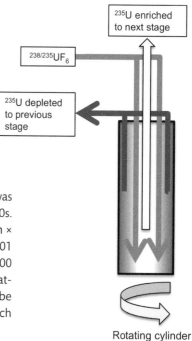

Many centrifuges are connected together in a 'cascade' arranged so that enriched UF_6 from one stage is passed to the next stage for further enrichment, and depleted UF_6 is returned to a previous stage. Enrichment from natural abundance to 3.5% for fuel use requires 9 stages, and enrichment to weapons grade (90%) can be achieved in 27 stages. The main disadvantage of gas centrifuge technology is small throughput, which is of the order of mg per s for an individual centrifuge, and so many centrifuges must be linked in a 'cascade'. A plant containing 36 × 164-machine cascades can produce up to 90 kg of weapons-grade U in a year. The high energy efficiency of gas centrifugation compensates for small throughput.

7.4.3 Laser separation

Due to mass differences, $^{238}UF_6$ and $^{235}UF_6$ have slightly different vibrational frequencies (the difference in ν_3 between $^{235}UF_6$ and $^{238}UF_6$ is 0.6 cm^{-1}) that can be selectively excited by laser (particularly at low temperatures), and this has been investigated for many years as a potential route to isotopic enrichment. The first proposed method, Molecular Laser Isotope Separation (MLIS), worked by selective excitation of a ^{235}U–F stretch resulting in dissociation of F and formation of involatile $^{235}UF_5$ which could be easily separated from gas phase $^{235/238}UF_6$. This process required extremely high energy input and so has been abandoned.

Separation of Isotopes by Laser Excitation (SILEX) is a more promising laser process that was developed during the 1990s. SILEX uses selective excitation of the ν_3 stretching mode of $^{235}UF_6$ (628 cm^{-1}). $^{235/238}UF_6$ diluted in a carrier gas (e.g. SF_6) at a concentration of <5% is expanded through a nozzle into a low-pressure chamber to form a supersonic beam. Within this beam, UF_6 and carrier gas molecules can interact at low temperature through van der Waals interactions to form 'dimers'. Selective excitation of the ν_3 stretching mode of $^{235}UF_6$ causes the $^{235}UF_6$ 'dimers' to dissociate very rapidly, and $^{235}UF_6$ molecules thus acquire kinetic energy, allowing them to migrate to the outer edge of the chamber from which they can be collected. Vibrationally non-excited $^{238}UF_6$ 'dimers' continue to condense to form particles. A schematic diagram of the SILEX process is shown in Figure 7.10.

There are serious concerns that the SILEX process could pose a real threat to nuclear arms proliferation: both space required and the energy consumption for a plant using SILEX technology are estimated to be much less than required for an equivalent gas-centrifuge plant. SILEX technology is not currently (2018) used commercially but is projected to come into small-scale production by 2020.

7.4.4 Energy costs of isotopic enrichment

Work must be done to achieve isotope separation as the process results in a decrease in entropy. The work done in isotopic enrichment is quantified in terms of 'separative work units' (SWUs). The number of SWUs required depends on the isotopic composition of the feedstock, the product, and the residue (or 'tailings'). Table 7.1 shows SWUs required starting from natural abundance uranium (0.711% ^{235}U).

The typical energy cost per SWU depends on the separation method. Gas diffusion has the highest energy costs and so is no longer used.

Energy costs per SWU

Method	Approx energy per SWU/kWh
Gas diffusion	2500
Gas centrifuge	40
Laser	10–15

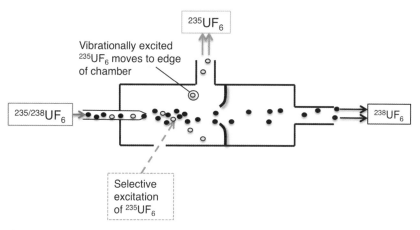

Figure 7.10 Schematic diagram of SILEX process

Table 7.1 SWUs required to obtain various levels of ^{235}U enrichment

% ^{235}U in enriched product	% ^{235}U in tailings	mass feedstock UF_6 to produce 1 kg enriched U	SWU required to produce 1 kg enriched U
5	0.25	10.3 kg	7.92
5	0.20	9.39 kg	8.85
90	0.25	194.7 kg	208

7.5 Recycling of spent nuclear fuel

As time progresses, the ^{235}U content of nuclear fuel rods becomes depleted, and there is a build-up of fission products, minor actinoids, and plutonium. Beyond a certain point, an efficient nuclear reaction can no longer be maintained—the concentration of ^{235}U is too low, and some of the fission products have a high neutron capture cross-section—and the fuel rod needs to be replaced. On average, one-third of nuclear fuel rods in a reactor are replaced approximately every 18 months. The spent fuel still contains approximately 1% ^{235}U and up to 0.6% fissionable ^{239}Pu, and this can be recycled into new fuel, while separating fission products and minor actinoids for safe storage. Due to the radioactivity of fission products, spent fuel rods can still generate a significant amount of heat, and so they are normally stored in a cooling pond for at least a year prior to reprocessing.

The most widely used procedure for recycling nuclear fuels is the PUREX (Plutonium and Uranium Recovery by EXtraction) process, which has been used since the 1950s (Figure 7.11). The key outcomes of the process are:

- Separation of U and Pu from fission products and minor actinoids.
- Separation of U from Pu.
- Safe storage of radioactive fission products and minor actinoids.

The PUREX process makes use of well-established coordination and redox chemistry of actinoids. The first step of the process is solvent extraction of uranium and

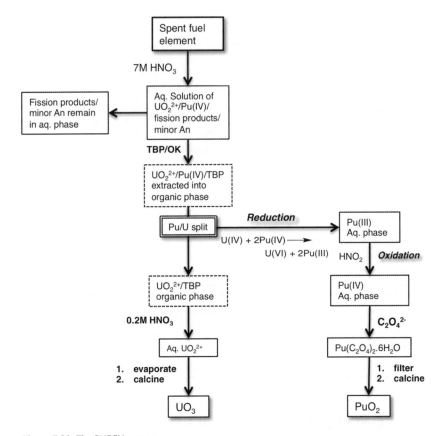

Figure 7.11 The PUREX process

plutonium from fission products and minor actinoids, and it makes use of the affinity of the hard Lewis base $(Bu^nO)_3P=O$ for UO_2^{2+} and Pu(IV). Separation of uranium from plutonium (the Pu/U split) is achieved by reduction of Pu(IV) to Pu(III) by reaction with U(IV), which is oxidized to UO_2^{2+} in the process. Pu(III) no longer coordinates strongly to $(Bu^nO)_3P=O$ and so can be extracted back into the aqueous phase, while UO_2^{2+} remains in the organic phase.

7.5.1 Management of nuclear waste

Fissionable uranium and plutonium can be recycled into nuclear fuel, leaving the question of how to deal safely with the radioactive fission products and minor actinoids. Initial fission products arise from uneven splitting of ^{235}U and may have extremely short half-lives as shown in Figure 7.12.

After cooling for one year, the major radioactive fission products are ^{93}Zr ($t_{1/2}$ 1.5×10^6 y), ^{137}Cs ($t_{1/2}$ 30 y), ^{99}Tc ($t_{1/2}$ 2.13×10^5 y) and ^{90}Sr ($t_{1/2}$ 28.5 y).

Heavier actinoids (mainly Np, Am, and Cm) are formed by neutron capture as shown for example in Figure 7.13.

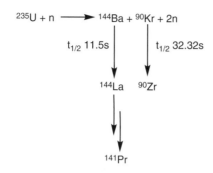

Figure 7.12 Initial products from ^{235}U fission

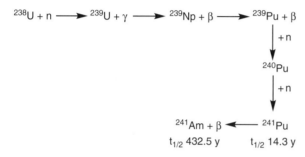

Figure 7.13 Neutron capture by ^{238}U to form heavier actinoids

The long half-lives of many fission products and actinoids mean that they must be stored for thousands of years before they become safe. This requires them to be in a solid form that is resistant to leaching; to achieve this they are incorporated into an inert matrix material such as ceramic or glass, and then sealed in stainless steel canisters for long-term storage.

The biggest challenge for dealing with nuclear waste is the extremely long time required for the waste to become safe. To address this problem, *partitioning-transmutation* is under consideration: the short- and long-lived radionuclides are first separated and then the long-lived species are converted by high-energy neutron bombardment into short-lived species.

7.6 **Summary**

The rare earth elements are by no means rare, but the isolation of the individual elements was a major challenge due to the similarity in their chemistry that results in their occurrence as mixtures in all of their ores. Fractional crystallization was the only method available until the development of ion exchange chromatography (a result of the Manhattan project during WW2). Industrial separation of the rare earths is now achieved by solvent extraction which, like ion exchange, exploits small differences of stability constants for complexation of Ln^{3+} ions with organic extractants as the Ln series is traversed. Th and U occur separately in nature, making their isolation much more straightforward than that of the lanthanoids. The most important aspects of actinoid separation are isotopic enrichment of natural uranium to the ^{235}U content required for nuclear reactors, and the processing of spent nuclear fuels.

7.7 Exercises

1. Bastnaesite ($LnFCO_3$) is an ore that contains mainly early Ln with the following relative composition:

Ln	La	Ce	Pr	Nd	Sm	Eu	Gd
%	33.2	49.1	4.34	12	0.8	0.1	0.2

   How much (i) La and (ii) Ce is extracted with 1 tonne of Nd? Compare these values with the data for rare earth consumption in Figure 6.1. What are the implications for the economics of rare earth production?

2. How is the lanthanoid contraction exploited in the separation of the lanthanoid elements?

3. Natural uranium consists of 99.3% ^{238}U and 0.7% ^{235}U. Consider a gas centrifuge using UF_6 with a separation factor of 1.25. Calculate the number of stages required to enrich from natural abundance to:

   (i) fuel grade (3.5% ^{235}U)

   (ii) weapons grade (90% ^{235}U)

7.8 Further reading

1. Binnemans, K., et al., 'Recycling of rare earths: a critical review'. *Journal of Cleaner Production*, 2013, **51**: 1–22.
2. Binnemans, K., et al., 'Rare earths and the balance problem: how to deal with changing markets?' *Journal of Sustainable Metallurgy*, 2018, **4**(1): 126–46.
3. Wilson, A.M., et al., 'Solvent extraction: the coordination chemistry behind extractive metallurgy'. *Chemical Society Reviews*, 2014, **43**(1): 123–34.
4. James, C., 'THULIUM I.1'. *Journal of the American Chemical Society*, 1911, **33**(8): 1332–44.
5. Wood, H.G., A. Glaser, and R.S. Kemp, 'The gas centrifuge and nuclear weapons proliferation'. *Physics Today*, 2008, **61**(9), 40.
6. Snyder, R., 'A proliferation assessment of third generation laser uranium enrichment technology'. *Science & Global Security*, 2016, **24**(2): 68–91.
7. Nash, K.L., et al., 'Actinide Separation Science and Technology', in *The Chemistry of the Actinide and Transactinide Elements*, L.R. Morss, N.M. Edelstein, and J. Fuger, Editors. 2011, Springer Netherlands: Dordrecht, 2622–798.

Glossary

Actinoid Any of the elements from Ac ($Z = 89$) to Lr ($Z = 103$). Abbreviated in this book as An. The term 'actinoid' is recommended by IUPAC; 'actinide' is often used for the elements Th to Lr.

Actinoid contraction The decrease in atomic and ionic radii with increasing Z from Ac to Lr.

Ancillary ligand An ancillary ligand (sometimes called a spectator ligand) does not take part in reactions of a heteroleptic complex. Its purpose is to provide a suitable steric and/or electronic environment to promote the desired reactivity in the reactive ligand(s).

Chelate effect The increased stability constant for a complex containing chelating (bi- or multidentate) ligands compared with that of an analogous complex containing monodentate ligands.

Coordination number The number of metal to ligand bonds in a complex, or the number of nearest neighbour atoms/ions in a solid-state structure.

Crystal field The electrostatic field that arises from the ligands surrounding a metal ion.

Crystal field splitting The splitting of degenerate electronic states by a crystal field. For most f-element complexes, the effect of the crystal field is smaller than that of spin-orbit coupling, and the result of crystal field splitting is to lift the degeneracy of $^{2S+1}L_J$ levels.

Hard Lewis acid/base A hard Lewis acid forms the most stable complexes with hard Lewis bases, and the bonding in these complexes is predominantly electrostatic. Hard Lewis acids and bases are both non-polarizable (i.e. their electron distributions are not easily distorted). Hard Lewis bases have very electronegative donor atoms (e.g. O or N).

Heteroleptic complex A complex in which the ligands are not all the same, e.g. [LaCp$_2$(CH$_2$SiMe$_3$)].

Homoleptic complex A complex in which all of the ligands are the same, e.g. [U(NEt$_2$)$_5$].

Hydrometallurgy An industrial process for separating/purifying metals by solvent extraction.

Inner sphere complex A central metal atom/ion together with the ligands that are directly bonded to it. The ligands are said to be in the inner coordination sphere.

Ionic radius An ion cannot have a precise radius due to the exponential fall-off of electron density with increasing distance from the nucleus. Ionic radius is therefore determined by apportioning cation and anion radii from numerous experimental measurements of interionic distances in solids. The radius for a particular ion increases with increasing coordination number, and it is therefore important to quote coordination number when using ionic radius data. The radii quoted in this book are based on a value of 140 pm for 6-coordinate O^{2-}.

IUPAC International Union of Pure and Applied Chemistry. Among other activities, it makes recommendations on chemical nomenclature.

Lanthanoid Any of the elements from La ($Z = 57$) to Lu ($Z = 71$). Abbreviated in this book as Ln. The term 'lanthanoid' is recommended by IUPAC; 'lanthanide' is often used for the elements Ce to Lu.

Lanthanoid contraction The decrease in atomic and ionic radii with increasing Z from La to Lu. (Eu and Yb atomic radii are anomalous.)

Laporte selection rule The Laporte selection rule states that for a transition to be allowed in electronic absorption spectroscopy, $\Delta L = \pm 1$. f$\rightarrow$f transitions are forbidden by the Laporte selection rule.

Luminescence Luminescence is a general term for the process of de-excitation by emission of photons rather than by non-radiative routes such as vibrational de-excitation. Fluorescence and phosphorescence are specific types of luminescence that are spin-allowed and spin-forbidden respectively.

Metallic radius Half of the experimentally determined nearest neighbour metal–metal distance in the solid.

Microstate A detailed description of the arrangement of electrons within a given electron configuration, e.g. for 4f^2: l = 3 m_l = 3 m_s = +½, l = 3 m_l = 3 m_s = −½; l = 3 m_l = 3 m_s = +½, l = 3 m_l = 2 m_s = +½ etc.

Outer sphere complex An outer sphere complex is formed by the electrostatic attraction between an inner sphere complex and other charged or uncharged species in solution. There is no direct bond between the central metal atom/ion and a ligand in the outer coordination sphere.

Phosphor A substance that displays luminescence.

Rare earth elements The lanthanoid elements plus scandium and yttrium. Often abbreviated as REE.

Rare earth magnet A powerful permanent magnet made from alloys containing lanthanoid elements (especially Nd, Sm, and Dy).

Relativistic effects Effects that must be considered in chemistry when electron speeds approach the speed of light. The effects are particularly pronounced for s-electrons of heavy elements (e.g. actinoids). Consequences of relativistic effects include lanthanoid and actinoid contractions, and destabilization of 5f and 6d orbitals of early actinoids.

Single ion magnet A special case of a single molecule magnet that contains just one metal ion in the complex.

Single molecule magnet A paramagnetic molecular complex that can retain magnetization in the absence of an external magnetic field. Often abbreviated as SMM.

Soft Lewis acid/base A soft Lewis acid forms the most stable complexes with soft Lewis bases, and the bonding in these complexes is predominantly covalent. Soft Lewis acids and bases are both polarizable (i.e. their electron distributions are easily distorted). Soft Lewis bases have less electronegative donor atoms (e.g. S or P).

Spin-orbit coupling Spin-orbit coupling is the interaction between the magnetic moment due to electron spin and the magnetic moment generated by orbital motion. The effect is to lift the degeneracy of ^{2S+1}L terms to produce $^{2S+1}L_J$ levels. The magnitude of spin-orbit coupling increases with increasing Z.

Stability constant Equilibrium constant for formation of a complex in solution, represented by the symbol β. Values are usually quoted as $\log_{10}\beta$.

Term symbol An abbreviated description of the angular momentum quantum numbers for a multi-electron atom or ion. It has the form $^{2S+1}L_J$. S is the spin quantum number, L represents the orbital angular momentum, and J is the total angular momentum quantum number.

Index